Praise for *The Rise of Wolf 8: Witnessing the Triumph of Yellowstone's Underdog*

"[Rick McIntyre's] greatest strength is the quiet respect and wonder with which he regards his subjects, a quality clearly informed by decades of careful watching."
PUBLISHERS WEEKLY

"Rick's book [*The Rise of Wolf 8*] is a goldmine for information on all aspects of wolf behavior and clearly shows they are clever, smart, and emotional beings."
PSYCHOLOGY TODAY

"The main attraction of this book, though, is the storytelling about individual wolves, including the powerful origin story of one of Yellowstone's greatest and most famous wolves."
WASHINGTON POST

"Yellowstone's resident wolf guru Rick McIntyre has been many things to many people: an expert tracker for the park's biologists, an indefatigable roadside interpreter for visitors, and an invaluable consultant to countless chronicles of the park's wolves—including my own. But he is first and foremost a storyteller whose encyclopedic knowledge of Yellowstone's wolf reintroduction project—now in its 25th year—is unparalleled."
NATE BLAKESLEE, *New York Times* best-selling author of *American Wolf*

"For many years I've thought that Rick McIntyre is the 'go-to guy' for all things wolf, and his latest book, *The Rise of Wolf 8*, amply confirms my belief. A must read—to which I'll return many times—for anyone interested in wolves and other nature. Wolves and humans are lucky to have Rick McIntyre."
MARC BEKOFF, PhD, author of *Rewilding Our Hearts* and *Canine Confidential*

"[McIntyre] spins the best stories about wolves that anyone will ever tell, ever. No one could match it."

DOUGLAS W. SMITH, senior wildlife biologist and project leader for the Yellowstone Gray Wolf Restoration Project

"*The Rise of Wolf 8* is a saga of triumph and tragedy, tribal warfare, love and loss, sagacity and survival, revealing new insights into the complexity of lupine existence. You will be richly rewarded by reading each detail patiently—they all lead to a gripping climax to an altogether stunning story."

NORMAN BISHOP, Yellowstone wolf interpreter and advisor to the Rocky Mountain Wolf Project and Living With Wolves

"This book is your invitation and opportunity to spend years inside Yellowstone National Park alongside the man who has spent more time watching wolves than anyone in the history of the world. As your patient teacher, he will show you that wolves are all individuals, and that the lives they lead are truly epic."

CARL SAFINA, author of *Beyond Words: What Animals Think and Feel*

"To follow the ever-changing destinies of the Yellowstone wolves is to witness a real-life drama, complete with acts of bravery, tragedy, sacrifice, and heroism. Rick McIntyre has monitored the park's wolves since reintroduction."

JIM AND JAMIE DUTCHER, founders of Living With Wolves

Praise for *The Reign of Wolf 21: The Saga of Yellowstone's Legendary Druid Pack*,
winner of the Reading the West Award for Best Narrative Nonfiction

---

"Wolf lovers, rejoice!"
BOOKLIST

"Like Thomas McNamee, David Mech, Barry Lopez, and other literary naturalists with an interest in wolf behavior, McIntyre writes with both elegance and flair, making complex biology and ethology a pleasure to read. Fans of wild wolves will eat this one up."
KIRKUS starred review

"Rick's passion for the Yellowstone wolves flows through this meticulous book about wolf love, play, life, and death. It's just like being there."
DR. DIANE BOYD, wolf biologist

"Rick McIntyre is a master storyteller and has dedicated his life to wolves—most particularly Yellowstone wolves. He tells their stories better than anyone, arguably better than anyone in history. I too have dedicated my life to wolves, yet reading Rick's stories, I still learn new things. This book is a treasure."
DOUGLAS W. SMITH, senior wildlife biologist and project leader for the Yellowstone Gray Wolf Restoration Project

"I'm always eager for the next book by Rick McIntyre. I learn so much fascinating information about wolves and their interactions with each other and with their prey."
L. DAVID MECH, author of *The Wolf: The Ecology and Behavior of an Endangered Species*

"Rick McIntyre has observed wild wolves more than any person ever. It is the way he sees wolves—as fellow social beings with stories to share—that makes his books so powerful. Through that lens, we glimpse our own hopes and dreams."

ED BANGS, former U.S. Fish and Wildlife Service wolf recovery coordinator for the Northern Rockies

"I was skeptical that McIntyre could write a second book as beguiling and insightful as his first about the wolves reintroduced to Yellowstone. Wow, was I wrong. This book is equally captivating as *The Rise of Wolf 8* (which you must read before *21*)."

KAY WOSEWICK, Boswell Book Company (Milwaukee, WI)

"What a tale! It reads like a well-written thriller, except it's real, and you'll never forget Wolf 21 after you've read his story."

LINDA BOND, Auntie's Bookstore (Spokane, WA)

"The continuation of the Yellowstone wolves series did not disappoint. The author did a fantastic job of telling the story of wolf 42 and 21 and the consequences and life of wild animals. I can't wait to see how this series will be concluded."

DAVID OTT, Books-A-Million

"Another fact-filled and science-based tale that can't help but also warm hearts."

SHELF AWARENESS

"[A] story of a fleeting, vibrant life that will change your mind not only about wolves, but about what a leader should do to project confidence and strength when the chips are down. From the first like to the last, it's a beauty."

ARKANSAS TIMES

Praise for *The Redemption of Wolf 302: From Renegade to Yellowstone Alpha Male*

"Rick's books should be mandatory reading for anyone who wants to learn about the fascinating animals with whom we share our magnificent planet and whose lives depend on our good will, decency, respect, along with a deep understanding and appreciation of how vital they are to maintaining a healthy planet in an increasingly human-dominated and troubled world."

MARC BEKOFF, PhD, author of *Rewilding Our Hearts* and *Canine Confidential*

"In this third saga tracing the lives of wolves in the wild, Rick rewards readers with a heartwarming, heartbreaking transition of Wolf 302 from a rambling rake to a responsible adult to a rearguard ready to die to save his pups. Wolf stories don't get better than this."

NORMAN BISHOP, Yellowstone wolf interpreter and advisor to the Rocky Mountain Wolf Project and Living With Wolves

"What thrilling stories about the Yellowstone wolves! The daily observations by Rick McIntyre are unprecedented in their detail and bring the personalities alive. The wolves are intelligent, feeling beings who, as exemplified by wolf 302, learn over a lifetime what their role in the pack is or ought to be."

FRANS DE WAAL, author of *Mama's Last Hug: Animal Emotions and What They Tell Us about Ourselves*

"Few people have ever observed wildlife as closely as Rick McIntyre, or written the biographies of individual animals with as much clarity and wisdom. *The Redemption of Wolf 302* opens yet another intimate window onto the lives of America's most beloved carnivores."

BEN GOLDFARB, author of *Eager: The Surprising, Secret Life of Beavers and Why They Matter*

"McIntyre's vivid accounts of Yellowstone wolves are reminiscent of Ernest Thompson Seton's animal tales, but these wolves are real and their skillfully told stories offer invaluable insights for researchers and readers alike."

BERND HEINRICH, professor emeritus of biology at the University of Vermont and author of *Mind of the Raven*

"What a gift to humanity is this latest account by Rick McIntyre, the world's most seasoned observer of wild wolves. The book is about birth and death, and living a full life from the perspective of another social species— and it is an unparalleled story of such a life."

ROLF O. PETERSON, biologist and author of *The Wolves of Isle Royale*

"Many books have been written 'about wolves.' Rick McIntyre writes about individuals whose lives follow the arc of singular careers. The frontier in understanding nonhuman animals is to see that they, like we, are individuals. In *The Redemption of Wolf 302* Rick McIntyre continues pushing further into the minds of individual free-living wolves he has known personally for many years. No one else's books have been written with the depth of time, personal insight, and sheer unstoppable devotion that Rick McIntyre brings to the page. The guy is the closest we have to a singular wolf who can write and tell stories."

CARL SAFINA, author of *Becoming Wild: How Animal Cultures Raise Families, Create Beauty, and Achieve Peace*

"Patient observer, kind soul, master storyteller, Rick McIntyre is one of a kind. Thousands who have met him in Yellowstone will attest to this. But he has done the most for the wolves; speaking for those who cannot, he has made their stories known. All right here in volume three of an amazing series of books about Yellowstone wolves. There has been nothing else like it."

DOUGLAS W. SMITH, senior wildlife biologist and project leader for the Yellowstone Gray Wolf Restoration Project

"Rick McIntyre's previous books describe wolf protagonists who shouldered responsibility for their packs with almost mythic courage. Here, in contrast, we meet a rebel who runs away from danger yet somehow always ends up with the girl. This engrossing tale of an especially intelligent and charismatic wolf reveals a deep and important lesson: for wolves as for humans there is more than one way to lead a good life."

BARBARA SMUTS, canine researcher and professor emerita at the University of Michigan

"Retired National Park ranger McIntyre continues his deeply revealing series on wolf behavior with this fine portrait of a lobo who makes good. . . . A great choice for anyone who has a fondness for wolves and an appreciation of good natural history."

KIRKUS REVIEWS

"With this third installment of Rick McIntyre's magnum opus, a new cast of characters brings fresh delights, while the cumulative impact of Rick's storytelling offers unexpected insights on wolves as social beings and on the culture of wolf packs. Yes, *culture*—read for yourself and see!"

NATE BLAKESLEE, author of *American Wolf*

Books in the
Alpha Wolves of Yellowstone Series
by Rick McIntyre

———————

The Rise of Wolf 8:
Witnessing the Triumph of
Yellowstone's Underdog

———————

The Reign of Wolf 21:
The Saga of Yellowstone's
Legendary Druid Pack

———————

The Redemption of Wolf 302:
From Renegade to
Yellowstone Alpha Male

———————

The Alpha Female Wolf:
The Fierce Legacy of
Yellowstone's 06

RICK MCINTYRE
Foreword by **JANE GOODALL**

THE
Alpha
Female
Wolf

THE FIERCE
LEGACY OF
YELLOWSTONE'S
06

GREYSTONE BOOKS
Vancouver/Berkeley/London

Greystone Books Ltd.
greystonebooks.com

Cataloguing data available from Library and Archives Canada
ISBN 978-1-77840-177-0 (pbk.)
ISBN 978-1-77164-858-5 (cloth)
ISBN 978-1-77164-859-2 (epub)

Editing by Jane Billinghurst
Copy editing by Brian Lynch
Proofreading by Meg Yamamoto
Maps by Kira Cassidy
Cover and text design by Fiona Siu
Cover photograph of the 06 Female by Robert L. Weselmann

Printed and bound in Canada on FSC® certified paper at Friesens. The FSC® label means that materials used for the product have been responsibly sourced.

This book was written after the author finished working for the National Park Service. Nothing in the writing is intended or should be interpreted as expressing or representing the official policy or positions of the U.S. government or any government departments or agencies.

Greystone Books thanks the Canada Council for the Arts, the British Columbia Arts Council, the Province of British Columbia through the Book Publishing Tax Credit, and the Government of Canada for supporting our publishing activities.

Greystone Books gratefully acknowledges the xʷməθkʷəy̓əm (Musqueam), Sḵwx̱wú7mesh (Squamish), and sə̓lílwətaʔɬ (Tsleil-Waututh) peoples on whose land our Vancouver head office is located.

CONTENTS

"Remember that hope is a good thing…
maybe the best of things, and no good thing ever dies."

STEPHEN KING, *RITA HAYWORTH*
AND SHAWSHANK REDEMPTION

"Hope itself is like a star—
not to be seen in the sunshine of prosperity,
and only to be discovered in the night of adversity."

C. H. SPURGEON, NINETEENTH-CENTURY
BRITISH PREACHER AND AUTHOR

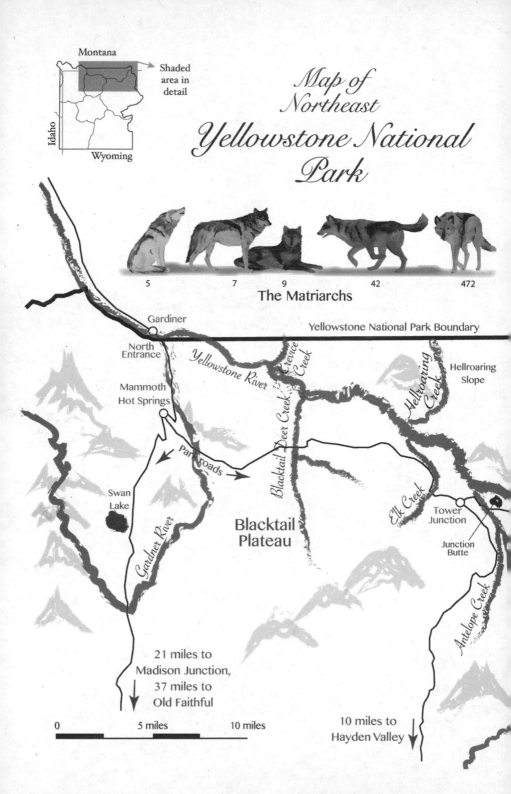

Montana

Idaho

Wyoming

Shaded
area in
detail

*Map of
Northeast
Yellowstone National
Park*

5   7   9   42   472

The Matriarchs

Gardiner

Yellowstone National Park Boundary

North
Entrance

Yellowstone River

Crevice Creek

Hellroaring Creek

Hellroaring
Slope

Mammoth
Hot Springs

Blacktail Deer Creek

Park roads

Swan
Lake

Gardner River

Blacktail
Plateau

Elk Creek

Tower
Junction

Junction
Butte

Antelope Creek

21 miles to
Madison Junction,
37 miles to
Old Faithful

0   5 miles   10 miles

10 miles to
Hayden Valley

Main Characters
2009–2015

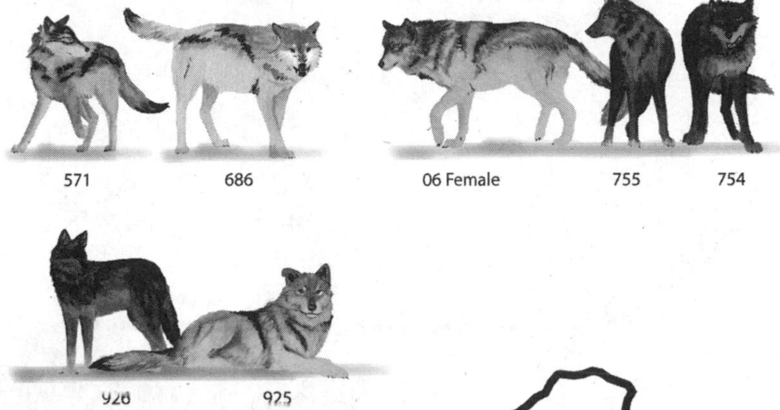

571   686   06 Female   755   754

928   925

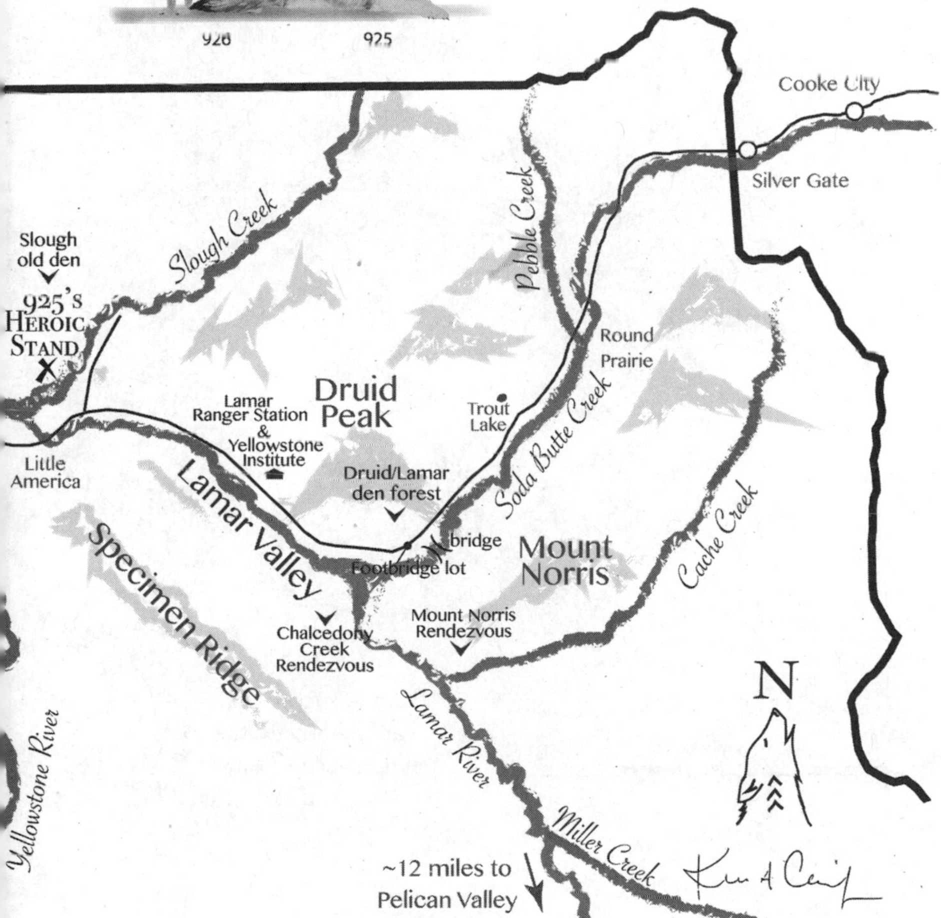

Slough old den

925's HEROIC STAND ✕

Little America

Slough Creek

Lamar Ranger Station & Yellowstone Institute

Druid Peak

Druid/Lamar den forest

Trout Lake

Pebble Creek

Round Prairie

Soda Butte Creek

Cooke City

Silver Gate

Cache Creek

Lamar Valley

Specimen Ridge

Footbridge lot

bridge

Chalcedony Creek Rendezvous

Mount Norris Rendezvous

Mount Norris

Lamar River

Yellowstone River

Miller Creek

~12 miles to Pelican Valley

N

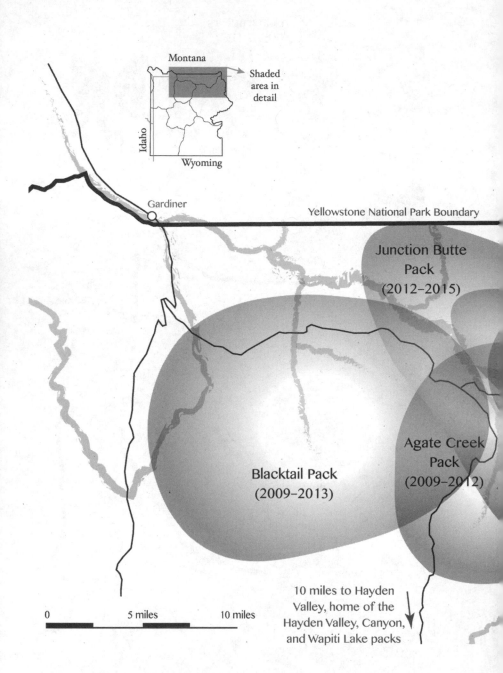

Montana

Shaded area in detail

Idaho

Wyoming

Gardiner

Yellowstone National Park Boundary

Junction Butte Pack
(2012–2015)

Blacktail Pack
(2009–2013)

Agate Creek Pack
(2009–2012)

0   5 miles   10 miles

10 miles to Hayden Valley, home of the Hayden Valley, Canyon, and Wapiti Lake packs

Select
Yellowstone Wolf Pack Territories 2009–2015

Cooke City

Lamar Canyon Pack
(2009–2015)

Silver Gate

Druid Peak Pack
(2009–2010)

Park road

~11 miles to
Hoodoo Creek, home
of the Hoodoo pack
→

N

~12 miles to
Pelican Valley, home
of the Mollie's pack ↓

THE MATRIARCHS

Wolf 5

As part of a reintroduction plan, wolf 5 was brought down to Yellowstone National Park from Canada in January 1995. Her family became known as the Crystal Creek pack, and they established a territory in Lamar Valley. The very next year, a new wolf family, the Druid Peak pack, was also relocated from Canada to the park. The Druids attacked the Crystal Creek pack, killing all of its members save for wolf 5 and her nephew wolf 6. Wolf 5 retreated south and settled into a new territory in a remote area, far from the Druid wolves. Her family was renamed the Mollie's pack in honor of Mollie Beattie, the first woman director of the U.S. Fish and Wildlife Service. Wolf 5's family thrived in that new territory. In later years, new generations of Mollie's wolves would make repeated incursions into Lamar Valley and carry on a long-standing feud with the Druid wolves. Mollie's alpha female 686, a descendant of wolf 5, would become an enemy of the 06 Female, but two other distant relatives of 686, brother wolves 754 and 755, were on 06's side when 686's pack attacked her family.

Wolf 7

Wolf 7 also arrived from Canada in January 1995 as part of the Rose Creek pack. As soon as the pack was released from its acclimation pen, she set off on her own and paired up with wolf 2 from the Crystal Creek pack. Together they founded the first naturally forming pack in Yellowstone since the 1920s, the Leopold pack, named after wildlife ecologist Aldo Leopold. Even though she was young and inexperienced, wolf 7 became a supremely successful matriarch. Her younger brother was the 06 Female's grandfather, wolf 21.

Wolf 9

Wolf 9 was the alpha female of the Rose Creek pack and mother of wolf 7. Shortly after her pack was released into Yellowstone in early 1995, wolf 9's partner, wolf 10, was illegally shot and killed outside the park. He died the same day 9 gave birth to a litter of eight pups. Her young family was returned to their acclimation pen. Wolf 8, son of Crystal Creek alpha female 5, later joined that family and adopted and helped raise 9's pups. One of her pups, wolf 21, was destined to become the most famous male wolf in Yellowstone history. Years later, after 9 was ousted from the pack by one of her daughters, she founded the Beartooth pack, just outside the park boundaries, a pack that continues to exist at the time of writing. Researchers believe that wolf 9 has more descendants in the Yellowstone area than any other female wolf. The 06 Female was wolf 9's great-granddaughter.

Wolf 42

Wolf 21 eventually left the Rose Creek pack and joined the pack that had driven the Crystal Creek pack out of Lamar: the Druids. The Druid alpha female, 40, was especially aggressive and drove her mother and a sister out of the family. She twice killed pups born to another sister, wolf 42. Wolves 21 and 40 were the parents of 472, the mother of the 06 Female. Soon after 472 was born, 40 was overthrown and mortally wounded by an alliance of females led by 42. That made 42 the new Druid alpha female. She raised her sister's pups alongside her own litter and reorganized the pack into a much better-run operation. 42 was the true leader of the pack and she, with the support of her devoted mate, alpha male 21, presided over what became a golden age for wolves in Yellowstone. Under 42's benevolent leadership, the pack grew to be the largest ever recorded anywhere in the world. At its peak, the Druid pack numbered thirty-eight wolves. 42 was the 06 Female's great-aunt.

Wolf 472

Druid alpha male 21 had many accomplished daughters who went on to found packs of their own. Perhaps the greatest was 472, the pup who had been born to 40 but was raised by 42. When she grew up, 472 left her family and eventually became the alpha female of the Agate Creek pack. Despite inheriting genes from her mother, 40, wolf 472 became a benevolent alpha female in the mold of 42, the wolf who had raised her. Her greatest accomplishment was training and mentoring her daughter, the 06 Female.

The 06 Female

The 06 Female, great-granddaughter of wolf 9, granddaughter of Druid Peak alphas 21 and 40, and daughter of Agate Creek alpha female 472, left her natal pack when she was very young to live as a lone wolf. She was courted by many suitors before she finally settled down and founded the Lamar Canyon Pack with two brothers, 754 and 755. These males were related to the Mollie's, the longtime enemies of the Druid Peak pack. 06's nemesis at the time of this story was Mollie's alpha female 686, who had the same type of violent personality as the 06 Female's grandmother wolf 40. To survive, 06 drew on the training she received from her mother, 472, who in turn had been mentored by the steady and wise wolf 42.

Wolf 926

One of the 06 Female's daughters, 926, took over the Lamar Canyon pack after the death of her mother. Her dramatic life story fully captures the resilience, resourcefulness, and strength that characterize all the female wolves in this book.

FOREWORD

RICK MCINTYRE IS the ultimate guru of wolf behavior—
he has been observing and documenting the behavior
of the wolves of Yellowstone National Park ever since
they were reintroduced in 1995. He started out as a seasonal
naturalist educating visitors about the wolves, but soon he
became a full-time observer of the packs.

Without question, Rick has spent more time watching
wolves than any other person. For an incredible fifteen-year
stint he did not miss one single day in the field. No matter
how he felt, no matter how bad the weather, Rick was out
there with the wolves. He knew them all as individuals and
followed many throughout their entire life. For animals with
big brains and complex societies, it is only long-term studies
of groups of known individuals that reveal the true complexity of their social life and document the range of behaviors
of which a species is capable. I started the study of chimpanzees in Gombe National Park in 1960 and the research
continues today, constantly revealing new behaviors.

Most people assume that I went to Africa wanting to work
with chimpanzees. But in fact, I was fascinated by all animals
from the earliest age. I watched the birds and squirrels and

insects around our house in the south of England, spending all the time I could out in nature and reading any books I could find about wild animals. (Television had not been invented back then!) I was ten years old when I decided, after reading *Tarzan of the Apes*, that I would grow up, live with wild animals, and write books about them. I did not want to be a scientist—just a naturalist—and I would have agreed to study any animal if it meant I could be out in the wild.

Although Africa was my first choice, there were many other countries I thought about and places I would love to have gone, places where there were vast areas that were (in the 1940s) truly wild. One of the animals that absolutely fascinated me was the wolf. Probably it was after reading Jack London's *The Call of the Wild* or maybe Rudyard Kipling's story of Mowgli, who was brought up by a pack of wolves in India. It just shows how important books can be in influencing the career choice of children, or at least instilling in them a love of nature. Of course, there are amazing documentaries about all manner of animals today, but there is something permanent about a book—it is there to be read and reread. I still have those books that influenced me as a child.

It was paleoanthropologist Dr. Louis Leakey who asked me if I would go and study the behavior of chimpanzees. As I slowly got to know the chimpanzees and their very different personalities and complex social relationships, I was struck by how much of their behavior resembles our own. Yet when Leakey got me enrolled to do a PhD at Cambridge University (even though I had never been to college!), many of the professors told me I had done everything wrong: I should not talk about chimpanzees as having personalities or minds capable of solving problems and certainly not as having emotions such as joy, sadness, anger, grief, and so on. These were attributes unique to humans, I was told. How fortunate that I had had a wise teacher throughout much of my childhood who had taught me that in this respect the professors were wrong. That teacher was my dog, Rusty. After all, dogs, like modern wolves, are descended from an ancient species of wolf!

In this book, Rick shares with us the drama of life in a wolf pack, just as I shared the drama of life in a chimpanzee community. As Rick discusses the rise and fall of these females through seven generations, the important role individuals play in the history of a pack becomes clear. One of the female wolves we learn about is 06 ("Oh-Six"). Rick writes that she was "a fiercely protective mother who fearlessly chased grizzly bears away from her pups and later had to repeatedly deal with a rival wolf pack that threatened to kill her family. 06's extraordinary life, along with stories of her daughter, wolf 926, and other greatly accomplished Yellowstone wolves, demonstrates the critical leadership roles females play in wolf society and their undaunted courage when facing threats and adversity." She reminds me of the old Gombe chimpanzee Flo, who was also fearless in defense of her offspring—and her daughter Fifi showed similar behavior.

How tragic that chimpanzees, wolves, and so many other amazing animals are threatened by human activities, habitat destruction, and hunting. Indeed, all life as we know it is threatened today by climate change and loss of biodiversity. Clearly it is desperately important that we work to reconnect people, especially children, with nature. For we are part of and not separate from the natural world. We are dependent on it for food, water, oxygen—everything. We must learn to understand and live in harmony with nature, with the chimpanzees of Gombe, the wolves of Yellowstone, and all the wild animals who live with us on Planet Earth. Each of us, in our own way, can help to heal the harm that humans have wrought.

A powerful way to touch people's hearts and change their minds is through stories, and Rick is a consummate storyteller. I hope the stories of the caring and resilient female wolves presented in this book will help people better understand the true nature of wolves, so that even the thousands who may never have the chance to see them in the wild will see them as Rick does: as loyal, playful, and smart, each one with a distinct personality, each one playing a role in the pack, each one a sentient being.

I would like to end on a personal note. Rick, by reading your books I have fulfilled one of my childhood dreams: your writing is so vivid, so powerful, that I feel I have been right there with you among the wolves of Yellowstone. And I urge you, the reader, to come with us and discover the magic of wolf society.

DR. JANE GOODALL, DBE, founder of the Jane Goodall Institute and UN Messenger of Peace

PROLOGUE

I KNEW THE 06 Female (pronounced "Oh-Six") from the time she was a pup through to the end of her life. I knew her parents, her four grandparents, and two of her great-grandparents. I also knew her sons and daughters and two generations beyond them. All that adds up to seven generations.

The 06 Female, so called because she was born in 2006, was a granddaughter of legendary Yellowstone wolf 21. Her parents were the alpha pair of the Agate Creek pack. She grew up to be a beautiful, free-spirited, independent wolf who had as many male suitors as a Disney princess. But 06 turned them all down for many years. Eventually she formed a pack with two brothers less than half her age. Soon after that, she had her first litter. 06 was a fiercely protective mother who fearlessly chased grizzly bears away from her pups and later had to repeatedly deal with a rival wolf pack that threatened to kill her family.

06's extraordinary life, along with stories of her daughter, wolf 926, and other greatly accomplished Yellowstone wolves, demonstrates the critical leadership roles females play in wolf society and their undaunted courage when facing threats and adversity.

THE HISTORY
OF THE DRUID PACK

T HE 06 FEMALE'S ancestors were members of what many people would consider the most famous wolf pack in the world. In 2003, the Druid Peak pack numbered thirty-eight wolves and controlled a vast territory in the northeastern section of Yellowstone National Park.

The longtime Druid alpha pair were wolf 21 and his mate 42. When 21 first joined the pack, the alpha female was 42's sister wolf 40. She used violence and aggression to dominate the other females in the pack. Evidence strongly suggested that for two years in a row, 40 killed 42's pups, probably to have all the pack's resources directed toward her own litters. The following year, 40 went to her sister's den for the likely purpose of once again killing her pups. That spring, 42 had several young female allies helping her and it appeared that in the process of protecting 42's pups, they ganged up on 40 and mortally wounded her. In her first act as the new alpha female, 42 carried her pups to 40's den, then got two other mother wolves to bring their pups to that central location. We eventually found out that 42 was raising her sister's pups along with her own litter. With the other two litters there, we had a count of twenty-one pups. Due to 42's leadership and organizational skills, twenty of those pups survived to the end of the year, an extraordinary success rate.

For the rest of her long life, 42 continued to be the Druid alpha female. Both she and 21 used cooperation over intimidation in their relationships with fellow pack members and exhibited empathy when one of their group got injured or was in distress. 21 was famous for never losing a fight with other male wolves and never killing a defeated opponent. Wolves 21 and 42 spent almost two-thirds of their lives together and had an exceptionally strong emotional bond. With wolf 42 leading the pack, 21 could concentrate on what he did best: going out on hunts, bringing back food to the family, and defending the Druids from rival packs. After 42 died in early 2004, 21 appeared depressed and listless. He eventually left his family, went up to a meadow where he and 42 had spent many days over the years, and died there.

By the time this story begins, five years after the deaths of 21 and 42, the once-mighty Druid pack had dwindled to just twelve members. The family's genetic legacy, however, was prospering owing to several daughters who had founded packs of their own in areas that had formerly been part of the Druid superterritory. One of these new packs was the Agate Creek pack, based just west of the Druid wolves. That pack, led by female 472, a daughter of 21, peacefully coexisted with their Druid neighbors. But the Hoodoo and Mollie's packs, unrelated wolves that lived in territories south and southeast of Lamar Valley, were more aggressive and posed a threat to the much-diminished Druid family.

PART I

2009–2010

1

Acting as the Alpha

THE FEMALE WOLF was desperately fighting for her life. Three big male wolves had pulled her down and were attacking her. She was on her back and snapping back at them as best she could. One of the males stood over her neck. He reached down and bit into her throat with all the force he could muster.

That gray female was 571, a three-and-a-half-year-old, radio-collared member of Yellowstone National Park's famous Druid Peak wolf pack. For many years, her family had dominated the park's northern region, but recently it had fallen on hard times. Their alpha female had been killed by rival wolves and the pack's two surviving pups and most of the ten adults had mange.

Mange is caused by mites that burrow into an animal's skin and feed on fluids and tissue. A wolf with mange incessantly scratches infected areas, which leads to missing patches of fur and open sores. At the time, mange was running rampant in Yellowstone and many packs were infected with it.

Mange causes tremendous suffering in wolves, especially when individuals lose most of their fur in the winter. What makes the issue even more terrible is something I came across in my research on the history of wolves in Montana. In 1905 the state legislature passed a law ordering the state veterinarian to capture wild wolves, infect them with the mange mites, then release them back into the wild so they could pass on the infestation to fellow family members. That bill was called "An Act to provide for the extermination of wolves and coyotes by inoculating the same with Mange."

Earlier that morning, 571 and her younger brother—a black yearling called Triangle, named for the shape of a blaze on his chest—had spotted the three male wolves in the Druid pack's rendezvous site. A rendezvous site is an area where a pack takes their pups when they are old enough to leave the den but not strong enough to travel with the older wolves when they go on hunts. The Druid rendezvous site, located at Chalcedony Creek, had been used by the pack for many years. It had water, a large open meadow, and an adjacent forest, all of which gave the pups plenty of places to explore and investigate.

Triangle and 571 watched as the three males howled close to where the family's two pups were last seen. Those males had recently left the Hoodoo pack, whose territory was just east of Yellowstone National Park. They were likely scouting out territory where they could start their own pack. The Hoodoos were the prime suspects in the recent death of the Druid alpha female. Seeing the rival wolves in the rendezvous site made me anxious, and I had to assume the two Druids felt the same way.

Both Druids howled, which gave away their position to the outsiders. The three males looked directly at Triangle and 571 as the two Druids continued to howl. Then 571, despite being far smaller in size, ran toward the rival wolves. Her instinct to protect the pups must have outweighed the high risk of being attacked and killed.

Triangle rushed after his sister. At that time, he could be described as the least impressive wolf in Yellowstone. He was very thin and suffering from an especially bad case of mange. He looked so pitiful that if he had been a dog in a shelter no one would have wanted to adopt him.

I watched as the three males charged at the two Druid wolves with their tails raised. They were moving away from where the pups were likely hiding, which was a good thing, but their vertical tails indicated they were intending to attack. When 571 and Triangle saw the males charging toward them, they turned around, ran back the way they came, then split up. 571 ran west and Triangle east, actions that confused the rival wolves. The Hoodoos hesitated, then all three ran after 571. She was now running for her life.

571 would have been a young adult when Triangle was born. Like other females of her age, she would have helped to feed and care for him and the rest of the pups in his litter. She likely spent a lot of time playing with him and his siblings. Her role in the pack was one of a caretaker to the younger pack members. As Triangle grew from pup to yearling, the two seemed to maintain a special attachment. This day, 571 had stayed behind with the family's pups when the rest of the pack had gone out on a hunt. It was up to her to protect them and Triangle from any threat. Since the three

males were now pursuing her, that meant she was doing her part to keep the younger wolves in her family safe.

As the three males chased 571, I saw that she was luring them farther and farther away from the pups. She was the fastest wolf in her family, so she had a good chance of out-running the males, but the lead Hoodoo male was fast too, and he was gaining on her. I realized the Druid wolf was heading toward the Lamar River and the nearby park road. If she could get across both of them, she would likely escape. The Hoodoo closed in and lunged. He bit into her rear end and pulled her down.

571 instantly jumped up, turned around, and attacked him. The male stepped back, startled by the ferocity of her defense. She immediately took off toward the water. He tackled her again but lost his balance and tripped. She got up first and got away. As she continued to run toward the river, he grabbed her for a third time and tossed her to the ground. She jumped up, turned around, snapped at his face, then ran off again.

When the rival wolf caught her for the fourth time, he was ready for her counterattack and dodged her attempted bite. A moment later, the other two males ran in. All three attacked 571, who was now on her back and biting up at them. One male was positioned over her belly and bit her there. Another was biting her hindquarters, and the third male, the biggest one, was standing over her head and shoulders. All three males were biting into her and violently shaking their heads, a tearing and ripping action intended to inflict serious damage.

The huge male looked down at her throat, then bit into it with all his strength. Normally that would be a fatal wound. But 571 refused to give up and continued to fight back against her attackers. What should have been a killing bite had done no damage. The wolf at her throat let go and appeared confused, acting like he did not understand why his bite had not killed her, but I knew what had happened. He had bitten into the front of her radio collar, the part that houses the electronics and battery. The collar had saved her life.

The attacking wolf quickly figured out the collar issue. He repositioned himself so he could bite into her throat from a different angle to avoid the collar. She had only a few moments to live before he made that bite.

I HAD KNOWN 571 since she was a pup and now was about to see her die. I was watching the fight from across the road and the image of her and the three males nearly filled the screen of my spotting scope. There was nothing I could do but record the last few moments of her life as she valiantly fought back.

Just as the large male was reaching down to make his fatal bite, I saw a skinny black wolf run into the scene. During the next few moments, the situation was too chaotic for me to figure out what was going on. Then I finally understood what was happening. Little Triangle had run back to save his big sister. He was darting back and forth like a movie action hero, attacking one male, then another.

Two of the males teamed up and charged at Triangle. He took off to the east. His surprise attack had ended the Hoodoos' assault on 571, and she was now running toward the river. But the big male was not ready to give up yet. He ignored Triangle and went after her. He caught her and resumed the attack. When the other males saw he had her, they ran over and joined him in biting 571. Right at that moment, Triangle ran back toward his sister. The attacking males saw him coming, stopped biting 571, and charged at her little brother.

571 jumped up and ran toward the water, which was now only a few yards away. I was hoping that this time she would make her escape. But the males looked back, turned around, and raced toward her. They got 571 again at the edge of the river. As she fought back, all four wolves ended up in the water, with the three males encircling the much smaller Druid female.

Triangle's courageous rescue attempt had only given his sister another few minutes of life. The Hoodoo males were now biting her at will. She had no hope of surviving the violent assault. At that moment, Triangle ran into the river at full speed and once again broke up the attack. He chased one of the males out of the water. The sight of such a small male chasing a much bigger wolf would have been comical if the situation had not been so dire.

571 used that distraction to run through the rest of the river, dash across the road, and quickly climb the slope on the other side. She ran right past me, then headed toward her family's den, a half mile to the east. A bloody gash across the width of her chest indicated she had been seriously wounded.

All three Hoodoo males were now chasing Triangle. One bit him on the hind leg, but Triangle yanked the leg free and continued running. Despite his poor physical condition, he succeeded in outracing his attackers. They slowed down, stopped, and watched him run farther away. Then they gave up and trotted off.

I later went over to the place where Triangle had broken free and found his tracks in the snow. Like 571, he had crossed the river and the road, then headed uphill. I heard from other wolf watchers that he had stopped and howled defiantly in the direction of the rival males, before limping toward the dense forest where the Druid pack had their den. I hoped he would meet up with his sister there and lick her many wounds.

Now that things had calmed down, I thought about 571. She had put a lot of time and effort into her relationship with Triangle, and the bond that she had with her younger brother saved her life that day.

Later that day, I found the radio-collared Druid alpha male, 480, and four other pack members a few miles west of where the Druid female had been attacked. They had been on a hunt and were returning to the den. By the next morning, radio-collar signals indicated that all the Druid wolves were in the forest that surrounded their den. The pack was back together, and Triangle and 571 were no longer the pups' lone defenders.

THE FOLLOWING DAY, the three Hoodoo wolves returned. I could not figure out why they were back. Did they intend to drive the Druids out of the valley and take over their territory? The Hoodoos howled repeatedly from across the road to the south and the Druids howled back from the north. 480, Triangle, and other Druids ran down to the road, crossed it, and charged at the intruders.

As I watched, I realized that the Druid alpha male was targeting the biggest of the Hoodoo males, the one that had nearly killed the Druid female. Rather than face the rapidly approaching Druids, the three Hoodoos turned tail and fled the area. They were spotted later on a mountain well to the south. The three males looked toward the den forest, then turned around, traveled over the top of the peak, and left the Druid territory.

In the days that followed the driving-off of the Hoodoo wolves, we saw Triangle traveling with the other adults and sometimes leading as the pack hunted. We had not seen 571 since she was wounded and reluctantly concluded that she had died from her injuries.

Early one morning, eleven days after the attack, I spotted the pack bedded down a few miles west of the den. I identified 480, Triangle, and some of the other adults. Other wolves were curled up and hard to see. After a while, a collared gray stood up, casually walked around, then went over to 480, greeted him, and pawed at his face. It was 571. She had survived and recovered from her wounds. The pack soon trotted off to the west with 571 in the lead. At one point, she had to stop and let the others catch up with her.

A few years earlier, 480 had defeated another, much larger pack when it invaded Druid territory. With the help of just a few pups, he had taken the initiative and charged at the invading force. 571 was one of those pups. When those three Hoodoo males came in, she saw a threat to her pack, took the initiative, and ran toward rival wolves. The main lesson I took from watching 571 that day is the critical role young females play in a wolf family. She could easily have saved herself by running away the moment she saw those three big males. But she was the only adult pack member there and had to step up. She took on the role of acting Druid alpha female and assumed all the responsibility of that position.

A Fierce Green Fire
in the Eyes of a Mother Wolf

When I saw 571 fighting back with such ferocious determination against those three big males, I thought of another female wolf who also was known for her ferocity, a wolf who radically changed a man's life, and how he, in turn, transformed America's attitudes toward wolves.

That man was Aldo Leopold, and in 1915 he was working for the U.S. Forest Service in northern New Mexico and was stationed on the Carson National Forest. Like other Forest Service employees, Leopold was told to kill any predators he spotted while out in the backcountry. That meant wolves as well as bears, mountain lions, and coyotes. Other federal employees were doing the same thing throughout the country at that time, including killing wolves inside Yellowstone National Park. That same year, Congress passed a bill funding a national program that had the stated purpose of "destroying wolves." The federal bureaucrat in charge of that program wrote that his agency's goal regarding wolves was "absolute extermination."

One day when out on a patrol with some coworkers, Leopold spotted seven wolves that turned out to be a family group. What he took to be an old mother wolf and her grown pups came together and had a greeting session. It would have been similar to scenes I have witnessed in Yellowstone thousands of times.

In an essay that Leopold later wrote, entitled "Thinking Like a Mountain," he describes what happened next.

> In those days we had never heard of passing up a chance to kill a wolf. In a second we were pumping lead into the pack, but with more excitement than accuracy.... When our rifles were empty, the old wolf was down, and a pup was dragging a leg into impassable slide-rocks.
>
> We reached the old wolf in time to watch a fierce green fire dying in her eyes. I realized then, and have known ever since, that there was something new to me in those eyes—something known only to her and to the mountain. I was young then, and full of trigger-itch; I thought that because fewer wolves meant more deer, that no wolves would mean hunters' paradise. But after seeing the green fire die, I sensed that neither the wolf nor the mountain agreed with such a view.
>
> Since then I have lived to see state after state extirpate its wolves. I have watched the face of many a newly wolfless mountain, and seen the south-facing slopes wrinkle with a maze of new deer trails. I have seen every edible bush and seedling browsed, first to anaemic desuetude, and then to death. I have seen every edible tree defoliated to the height of a saddlehorn.

Leopold later became a professor of wildlife management at the University of Wisconsin. Through his teaching there and in books such as *A Sand County Almanac*, which contained his "Thinking Like a Mountain" essay, he profoundly influenced generations of wildlife biologists to view wolves and other predators as essential members of a healthy wildlife population.

In late 1944, Leopold suggested something that seemed revolutionary at the time: wolves should be brought back to their former home in Yellowstone. Rangers had killed the last of the native park wolves in 1926, eleven years after Leopold shot the female wolf. In 1995, the National Park Service reintroduced wolves to the park and later named one of the new wolf families in Aldo's honor: the Leopold pack.

The impact Leopold had on the world might not have happened if he had never experienced that profound encounter with the mother wolf eighty years earlier. A bullet took her life, but the fire Leopold saw in her eyes, her life force, changed him. He then changed American attitudes toward wolves and inspired the eventual restoration of wolves to Yellowstone. She did not die in vain.

Bruce Babbitt, President Clinton's secretary of the interior, was one of the people who worked to restore wolves to Yellowstone. Here is what he said about the day the wolves arrived in the park:

In January 1995 I helped carry the first gray wolf into Yellowstone, where they had been eradicated by federal predator control policy only six decades earlier. Looking through the crate into her eyes, I reflected on how Aldo Leopold once took part in that policy, then eloquently challenged it. By illuminating for us how wolves play a critical role in the whole of creation, he expressed the ethic and laws which would later reintroduce them nearly a half-century after his death.

Since there was only one female in the first pack carried up to the acclimation pen that day, the wolf Babbitt spoke about must have been the Crystal Creek alpha female, wolf 5. She was a mother with pups, like the female Leopold shot. After the family was released, wolf 5 continued to have pups every spring for many years to come. Twenty-six years after Babbitt looked into her eyes, her descendants still control a prime-quality wolf territory in the central portion of Yellowstone.

At the beginning of "Thinking Like a Mountain," Leopold described hearing a wolf howl: "It is an outburst of wild defiant sorrow, and of contempt for all the adversities of the world." The full force of the United States government spent decades trying to exterminate wolves, but no matter how many they killed, some mother wolves survived in remote regions, such as the Lake States and Alaska, and kept on having pups. Thanks to their wild defiance and contempt for adversity, wolves were never fully exterminated in our country.

On the day she valiantly fought against those three big males, Yellowstone female 571 would have had the same defiant, fierce fire in her eyes that Leopold had seen nearly a century earlier.

2

Enter the
06 Female

I THOUGHT ABOUT THE former Druid alpha pair 21 and 42 in late October when I saw Druid alpha male 480 walking around the pack's rendezvous site. In my field notes for that day, I wrote: "480 seemed lethargic, like he was in a depression." His mate had died, and it looked like he was going through the same emotional experience that 21 did when he lost his longtime companion, wolf 42.

The February 2010 mating season was just a few months away. Wolves normally do not mate with family members, and that instinct helps maintain genetic diversity. All the other females in the pack were too closely related to 480 for him to breed with them, so I expected he would leave the family and try to find a new mate. For the time being, however, 480 remained in the role of alpha male. The nine younger Druid adults, seven females and two males, were also related to each other. What could save the Druid dynasty would be the arrival of an outsider male to breed with the new alpha female, known as White Line, and join the family.

On December 2, we spotted one of the Druid females alone under a tree. She was joined by a young black male with a mange-free coat. I was glad to see that he looked very healthy and wondered if this young wolf was destined to become the Druids' new alpha male. The newcomer and the Druid female stood face to face and acted friendly to each other. Another Druid female joined them, wagged her tail at the outlander, then playfully jumped around him. All of that showed they were very interested in this potential suitor. But how would the father of those females react to his arrival?

Then I saw 480 downhill from them. As he trotted up toward the three wolves, the new male moved off and met up with another one of 480's daughters. 480 ran at the black, and the two males confronted each other from just a few feet apart. The new wolf did not back down from the much larger alpha male, but he did lower his tail, signaling to the Druid alpha that he considered him the dominant wolf. The two wolves circled each other, then got in a fight. 480 tried to pounce on the black and pin him, but the younger male did not go down.

The females had been watching from the sidelines. One of them got between the battling males, and that calmed things down. I later thought that was a smart move, for the breeding season was rapidly approaching and the Druid sisters desperately needed to find an unrelated male. Here was a healthy suitor who seemed to solve that problem. They needed him to join the pack so the family could have pups in the spring. Their father's aggression was counterproductive to their agenda.

When the female that had broken up the fight walked off, the new black followed her. Later a fourth Druid sister came over and flirted with him. He had now met five of the Druids. After that, he got in a brief fight with Triangle, but two females stepped in between them and once again ended the aggression. The interactions between the newly arrived male and the Druids continued in a similar pattern for some time that day. The black male was later collared and designated 755, so I will refer to him that way from now on.

The following morning, we did not see 755 near the Druids, but several of the females were missing and I figured they were with him. On December 4, we found the Druids on the north side of Lamar Valley. 755 was with five of the pack's females. Later 480 joined them and repeatedly chased 755, but the younger male persisted in staying in the area and always got back with the females.

480 continued to chase 755 away for a few more days. Then I saw that 755 was doing scent marking when he was among the Druid females, a sign he was now part of the pack. I felt the Druid females had made the decision to incorporate 755 into the family despite their father's aggression toward him. Now 480 was going to have to make his own decision: either stay with the Druids and accept that reality or take off and try to find a new mate.

THIS WAS MY eleventh winter in Yellowstone and I still was struggling to deal with the frigid weather. December 2009 turned out to be especially cold. On the tenth, the low was minus 22 Fahrenheit (–30 Celsius). I had four layers of clothes on my legs, eleven on my upper body, and was still cold. I had a great deal of sympathy for the wolves who had lost sections of their coats to mange. 480 was still in the pack at that time and continued to drive off 755, but the persistent young male always got back with the females.

Two days later, I saw the 06 Female. She had a strikingly beautiful gray coat. She had not been collared and so she had no official number. Her informal name had been coined by longtime wolf watcher Laurie Lyman to separate her from a near-identical sister born the following year. That sister became known as the 07 Female. 06 had grown up in the Agate pack and was a granddaughter to Druid 21 and one-time alpha female wolf 40. When I spotted her, she was with one of her sisters and an unrelated male, but 06 also often traveled as a lone wolf, a category that in Yellowstone comprises only about 2 percent of the wolf population.

Most lone wolves are males who have left home and are looking to join a new pack or to find a female to start a pack of their own. Wolves have a matrilineal society, meaning a female leads the pack, and when she dies, one of her adult daughters usually inherits the leadership position. Because wolves are averse to mating with relatives, that system usually causes many of the sons to leave the family. They become lone wolves as they try to seek out a female from another pack. Lone females are much less common. 06 had been courted by numerous big males during the previous few years but had yet to settle down and start a family. I was beginning to think she was too independent and self-sufficient to pair off with a male. However, the truth was likely that she had just not met the right wolf yet.

I lost sight of the trio, then picked up 06 chasing a cow elk by herself. As she closed in, the cow stumbled and the wolf bit into her. They ran out of sight with 06 about to attack her again. I went to a different angle and saw that 06 had killed the elk. She was sharing the meat with the two other wolves. 06 was a master hunter, and the hunt we had just witnessed was proof of her superior abilities.

After that, things got complicated. The Druids ran in and drove off 06 and her two companions. The group of Druids included 480 and 755, along with many females. That meant that the two males were briefly in an alliance to take the kill away from the smaller pack. After eating, 755 flirted with 480's daughters. I noticed that White Line was doing a lot of scent marking, which confirmed she was the alpha female of the group. 755 went there, sniffed her scent marking, and marked over it. I took that to mean that he now saw himself as the pack's alpha male. I saw 06 standing on a nearby high butte. She was watching the Druids. That likely was her first sighting of 755.

THE NEXT DAY, December 13, the Druids and 06's group, which we called the Lava Creek pack, were still in the area. But I was not getting any signal from 480. All the other Druids were there, along with 755. The two packs howled at each other. The Druids had left the elk carcass, so the Lava wolves went back to it and fed. They also sniffed around and would have picked up the scent of the other wolves. Bob Wayne, a professor of ecology and evolutionary biology at the University of California, Los Angeles, once told me that wolves and dogs can distinguish between forty thousand and fifty thousand different scents. That is how a wolf can sniff at a track and identify other individual wolves and prey animals. 06 would have investigated all the scents around that carcass and therefore would have had what for a wolf would be an introduction to the male we knew as 755.

After moving off, 06 looked toward the Druids. She howled and 755 howled back, then the rest of the Druids joined in. 06 howled again and 755 did another answering call. She howled in response. Those howling duets between 06 and 755 turned out to be a sign of what was to come.

IN LATE MORNING, I found 480 a few miles away from the rest of the Druids. His family howled. He must have heard them but did not howl back. 480 was looking away from the Druids. He moved out of sight after that. I took that to mean he had made his decision to leave his family and set out to find a mate.

Two days later, I spotted 06's group at Slough Creek. The main Druid group howled from the north and the Lava wolves howled back, with 06 doing the most howling. 06 led her group toward the sound of the Druid howls. She spotted one of the Druid females and chased her. 06 and her sister, 471, caught up with the Druid and had a standoff with her. The females snapped at each other but did not get any real bites in. 06 and 471 were related to all the Druid females, so perhaps that mitigated the aggression. The Lava male arrived but did not get involved in the dispute among the females, always a smart thing for a male to do.

I FREQUENTLY GOT 480's signal away from the main Druid group during the coming days and occasionally saw him. He was always alone and spent a lot of time bedded down, more than normal. That could have meant he was in declining health. I saw him resting in a snow-free spot under a tree on December 21, and he was still there two days later. Eventually he got up, fed on an old carcass, then bedded down once more.

Two more days passed, and we found that 480 had moved west to the area around Hellroaring Creek. He was close to a carcass but moving away from it. 755 and the rest of the Druids were nearby, and they had probably made the kill. Later I saw 480 go back to that kill and feed. 755 walked over to the carcass. The two males got close to each other, and 480 lunged and snapped at 755. That happened a few more times, but I noticed that 755 was not aggressive to the older male and did not react to the lunges. Soon three Druid females arrived and fed beside their father as 755 bedded down nearby.

The older male eventually walked off to the west, and I soon lost sight of him. That would be the last time I saw 480. I later thought a lot about how 755 treated 480 that day. I think he respected 480 and chose not to bother him. The incident reminded me of the times I had seen alpha males like 21 and 480 spare the life of a defeated rival male. 755 seemed to have the same type of personality as those great Druid males.

As I thought over those recent events, I saw that the females in the Druid pack had ultimately controlled the situation. Their agenda to bring in an unrelated male so they could breed and have pups the next spring had won out over their father, the pack's alpha male, and his desire to stay in the family. It now looked like the Druid pack would be revitalized with the addition of 755. However, sometimes agendas do not work out as planned. The Druid genetic line would flourish, but not in the way we expected.

3

06 Puts Together a Team

IN EARLY JANUARY 2010, we continued to see the 06 Female with her two companions, and they often were near the Druids. She did a lot of howling as she looked toward them. Now that 480 was gone, 755 was the new Druid alpha male. Triangle was still in the group and 755 seemed to be treating him well. I saw Triangle licking a mangy section of one of his sisters' coats. That may have relieved some of her discomfort there. Soon after that, Triangle left the group and went west by himself, possibly to look for his father, 480.

The winter weather was getting much colder. The official low at the Lamar Ranger Station on January 7 was minus 43 Fahrenheit (–42 Celsius). That was the coldest temperature yet in my time in the park.

I spotted the Druid wolves again on January 14 and noticed that a new black was with the group, another male with a healthy coat. He went to 755 in a submissive crouch and greeted him. After watching the two males interact that day, I concluded that they were likely brothers who had left their original family, gotten separated, then just reunited. The new male was later collared and given the number 754. He always acted subordinate to 755, despite being bigger. By that time, there were five females in the Druid group, along with the two new males. We had lost track of the other pack members, including Triangle and 571, the female who had risked her life to save the Druid pups.

LATE IN JANUARY, I got a report of six Druids chasing an uncollared gray. I saw 06 there and thought she must have been the one they had chased. After that, I spotted her with 754 and they were acting friendly to each other. The next morning, both 754 and 755 were bedded down with 06 and none of the Druid females were nearby. 06 got up and walked off. The two males followed her.

06 was strong and healthy, as well as a good hunter. The Druid females all had mange, were in poor condition, and often seemed listless. As I saw the hairless patches on their bodies, I worried about their ability to survive the extremely cold weather we were having. My former Yellowstone National Park Wolf Project colleague Emily Almberg and a team of researchers had studied that issue. Using thermal imagery, they found that wolves with severe mange lose 65 to 78 percent more body heat on winter nights than do wolves with full coats, coats that insulate them from the extremely cold temperatures we experience in Yellowstone. The researchers estimated that a wolf with severe mange has

to consume 1,700 more calories per day to compensate for that heat loss.

I could understand why 754 and 755 were drawn to the strong and healthy 06. The brothers did a lot of scent marking and howling with her, a sign the three wolves were bonding and forming a new pack. By late January, I could see that 06 was coming into breeding condition and the brothers were obviously very interested in her. It seemed she had chosen them over her sister and their male companion, and the brothers had chosen her over the remaining Druid females.

754 and 755 were darted and collared around that time. They were both classified as yearlings and each weighed around 100 pounds. 06 was slightly smaller—when she was later collared, she weighed around 94 pounds. When 06 met the two males, she was nearly four, meaning she was twice their age. When I gave talks to park visitors about 06 and the two males, and mentioned that age difference, nearly always a woman in the group would yell out, "She's a cougar!"

By that time, I had been working for the Wolf Project for twelve years, helping gather data on packs in the park. One of our staff biologists, Dan Stahler, told me that blood samples taken from the two males when they were collared found that they were brothers, as we thought. Their mother had been born into the Nez Perce pack on the west side of the park but left her family to start a new pack near Old Faithful. Their aunt, wolf 486, also left Nez Perce and later became the alpha female in the Mollie's pack. That meant the brothers were related to the Mollie's wolves. The Mollie's lived south of the Druids, and the rivalry between the two packs stretched back to the time the original Druid wolves arrived in Yellowstone from British Columbia, Canada, fourteen years earlier.

I had known 06's father, Agate alpha male 113, for many years. He had a calm and confident personality, avoided unnecessary battles with neighboring packs, and was totally devoted to his family. 113 was a nephew to 21 and had a character much like his. I guessed that 06 would have seen her father as the model of an alpha male and discerned those same qualities in 755. We would eventually see him turn into a classic alpha male who behaved much like 113. Likewise, 754 proved himself to be a superior wolf many times in the coming years. 06 knew what she was doing when she picked those brothers.

A LITTLE WHILE later, 06 was temporarily away from the brothers. Big Blaze, a former Druid male who now was the alpha male in the Agate pack, approached her wagging his tail and looking friendly. When he reached down to sniff her, she reacted by grabbing his throat. That put him at her mercy, for if she used the great strength in her jaws, she could kill him. He wisely just stood there and did not struggle or panic. 06 let him go but grabbed him by the side of the neck when he approached her rear end. After that he tried once more, and she bit him hard enough to make him squeal in pain.

Big Blaze had mating on his mind, but it was clear that 06 wanted nothing to do with him. 06 was a tough female and would not put up with any annoyance from a male, even if he was much bigger than she was. When watching her easily dominate Big Blaze, I thought 06 could have been an Amazon warrior reborn as a wolf. Her reaction to that male was the confirmation she had chosen to settle down with 754 and 755. 06 had picked her teammates for her new pack and was going to stick with them. That decision was likely going to be the most important one of her life, for her fate—along with the fates of any pups she might have—depended on those males living up to the qualities she must have seen in them.

On the tenth of February, 06 was south of the park road, and one of the black males, either 754 or 755, was to the north. She wanted to cross the pavement and go to him, but a large crowd had gathered and people trying to get photos were blocking her. 06 quickly trotted east to get around them. About twenty people ran along the road, hoping to get more shots of her. I was on the scene, trying to get everyone to stop, but too many were well ahead of me, running after her. Then Tom Murphy, a local photographer who had stayed in place, had enough. In a loud voice he yelled out, "Why don't you let her cross the road?" Everyone stopped, and 06 crossed and headed north.

She was soon back with the two black males, and I saw 755 licking 06's face. Unlike her aggressive response to Big Blaze, she accepted 755's affectionate attentions. That month, 06 mated with both brothers.

By then, we had lost track of most of the Druid wolves. Only White Line and one other female were still together. I saw them attack and kill an adult coyote, and then White Line did something I had never seen a wolf do: she ate part of that full-grown coyote. I thought her poor health and weight loss had made her desperate for any form of meat. Her sister did not touch the coyote. In late February, we found the remains of White Line. It looked like she had been killed by a mountain lion. With no sightings of any other Druid wolves, we reluctantly concluded that the pack had disbanded.

BY LATE MARCH, 06 looked pregnant. We had seen her near the Slough Creek pack's old den site and it looked like she was going to have her pups there. The Slough Creek pack had been founded by one of 21's daughters. Starting in 2002, it had occupied a territory north of the Agates and west of the Druids. By 2010, the few remaining Slough females were suffering from mange, and the pack soon ceased to exist. Their old den was located on an open ridge a mile west of the creek.

That site was ideal for 06, for there were many burrows nearby where the pups could hide in times of danger. There was water close to the den and good hunting opportunities in the surrounding countryside. It was also ideal for us, as we could see the whole area from our viewpoint across the creek, about a mile from the den. There was only one downside: a lot of grizzlies passed through the area, and 06 would likely have to deal with them. The Wolf Project decided to name 06's new group the Lamar Canyon pack, based on that nearby landmark.

Since 06 came from the Druid genetic lineage, her new family would be a continuation of that dynasty. A granddaughter of wolf 21 had become the alpha female of this revitalized version of the pack that he had made world-famous. Many park visitors thought of 06 as a princess. Now she was the queen of what had once been the Druid territory.

The Founding Mothers

Having 06 and her family den near Slough Creek brought back a lot of memories of two other packs that lived in that general area in the early years of Yellowstone's wolf reintroduction program. Those wolf families had greatly accomplished alpha females like 06.

The Crystal Creek and Rose Creek packs were brought down from Alberta for the reintroduction in January 1995. The Crystal wolves were released from their acclimation pen south of Slough Creek and later settled in Lamar Valley.

The original Rose Creek alpha male was illegally shot and killed outside the park soon after his pack was set free. That made alpha female 9 a single mother with eight newborn pups. National Park Service biologists captured the family, put them back in their original pen, and released them six months later.

Mother wolf 9 was desperate for help with her big litter and needed an adult male to join her family. Yearling wolf 8 from the Crystal Creek pack came along, befriended some of the pups, then was recruited into the Rose pack by 9 as the new alpha male. With the Crystal wolves occupying Lamar Valley, the Rose wolves established their territory a few miles to the west in the Slough Creek area, where 06 later denned in 2010. In both cases, it was the alpha female who decided where her family would settle and where her pups would be born.

In the spring of 1996, the Crystal alpha female, wolf
5, denned near what is now known as the den forest in
Lamar Valley. The Druid pack, a family brought down
in the second year of the reintroduction, was released
around that time. Wanting Lamar for themselves, they
attacked the Crystal wolves at that den site and killed
their alpha male. None of the Crystal newborn pups
survived the incident. The only pack members left alive
were wolf 5 and a young male, wolf 6. Since the Druids
now controlled Lamar Valley, the two Crystal wolves
left and established a new territory in a remote region
known as Pelican Valley, well south of Lamar Valley.

Wolf 9's recruitment of 8 into her family turned
out to be a masterful decision, for he became one of
our greatest alpha males. But 9 was always the leader
of the pack.

Wolf 9 continued to have tragedies and losses in
her life. When she was an old wolf, an adult daughter
turned on her and took over the alpha female position.
Rather than accept a subordinate role in the family, 9
took off and established a pack east of Yellowstone. It
became known as the Beartooth pack and is still in
existence today, decades after she founded it. As far
as we can determine, wolf 9 has more descendants in
the Yellowstone area than any other female wolf. Wolf
Project records show that 9 had at least thirty pups
during her years in Yellowstone, making her one of the
most important founding females of the region.

Like 9, wolf 5 faced a long series of challenges
as she led her pack over the years. Her pack had

specialized in elk hunting, but deep winter snow caused the elk in their new territory to migrate to lower elevations with more moderate weather. Only bison remained in Pelican Valley during the winter, so 5 and her pack had to learn how to hunt and kill animals that were up to twenty times bigger than an average wolf. They figured that out and thrived in the territory.

The current generation of this family is the only one of the original reintroduced packs still in existence as I write this book, twenty-seven years after wolves were brought back to Yellowstone. That is a testament to 5's decision to choose that valley as the pack's territory after they had been driven out of Lamar Valley by the Druids. The pack, now known as the Mollie's pack, was renamed in honor of Mollie Beattie, the first woman director of the U.S. Fish and Wildlife Service, who was a fierce advocate of wolf restoration in the West.

Regardless of the setbacks those two early alpha females suffered, they continued on with their lives and agendas, always moving forward. They found ways to overcome whatever hard times they encountered. My friend Laurie Lyman, who knows the Yellowstone female wolves so well, put it best: "For those females, giving up was not an option."

4

06 and Her First Pups

B Y EARLY APRIL, there was a well-worn trail through the snow to the entrance to the den at Slough Creek. In the middle of the month, I watched as 754 and 755 approached the den opening. One of them wagged his tail, slipped into the den, then backed out. 06 came out a moment later and greeted the males. I could see that she was still pregnant. She bedded down with them for a couple of hours before going back into the den.

755 and 754 had already begun the process of provisioning 06 at her den. They would fill their stomachs at a kill site with as much as twenty pounds of meat, then return to the den. When they arrived, 06 would rush out of the den and frantically lick their muzzles. That licking triggered a regurgitation of the partially digested meat and she would quickly gulp it down. For wolves, that is a far more efficient method

of transporting meat than carrying it in their mouths. When pups are old enough to eat solid food, they mob incoming adults and leap up to lick their mouths, begging for a feeding. Another advantage of that food delivery system is that it spares a mother wolf from the dangers of hunting big prey animals that might injure or kill her, and enables her to concentrate on caring for the pups.

We saw a typical example of that feeding process in late April. 755 was spotted feeding on a carcass at Hellroaring. He left the site at 7:20 that morning and headed east. I saw him arrive at the den three hours later and regurgitate meat to 06. The straight-line distance from the carcass to the den would have been about twelve miles, but that included a lot of up-and-down travel over rough terrain.

06 sometimes demonstrated how she could outsmart her two male companions if one of them tried to save some of the meat for himself. One day I saw 754 bury meat in two separate places near the den, hiding the pieces so he could come back later and eat them. Less than an hour after that, I spotted 06 sniffing around one of those sites. She must have found the cache, for I could see her eating there. When she finished, 06 walked off and seemed to be searching for the other cache. She moved back and forth, soon found that second site, and ate the hidden meat. 06 was an expert at finding and following scent trails and used that skill to foil 754's plans.

By that time, 06 no longer looked pregnant, so she must have had her pups. She had never been a mother before, so we were eager to see how she cared for the litter. There were other signs of spring in Yellowstone. Green grass was coming in along the road, a welcome sight after the long winter. On May 1, I saw the first flowers of the year.

A GRIZZLY MOTHER with two yearling cubs showed up near 06's den on May 9. The mother wolf ran at them with her tail raised and tried to nip one of the bears. The sow charged at the wolf and drove her off, but 06 came right back and snapped at the mother bear's rear end. That sequence was repeated many times. I then realized what 06 was doing. The den and her pups were a short distance northwest of the bears. The wolf would come in from the opposite direction and run off back that way when the bears charged her. She was cleverly luring them farther and farther away from the den, like a mother bird pretending to have a broken wing.

Local wolf watcher Doug McLaughlin was on the scene before me and told me that one of the yearling bears had approached the den. When it was about ten feet from the entrance, 06 charged out of the tunnel and attacked the young bear. As they fought, the two opponents moved downhill, away from the den. Doug told me 06 bit the bear several times. The mother grizzly ran in and tried to protect her cub, but 06 continued to bite at it.

Then the wolf turned her attention to the cub's mother. She lunged at the bear but stopped just before reaching her. The sow chased her in a direction away from the den. I felt that was exactly what 06 wanted her to do. 06 ran back and confronted a yearling. It charged at her and swung a front paw at the mother wolf. I could hear 06 vocalizing at the bear family, a high-pitched call with a barking component. That sound would carry a long distance and alert the two males to the crisis at the den.

06 was taunting the bears, annoying and frustrating them so they would forget about the den and concentrate on trying to catch and kill her. She continued to lure them farther

away from her pups. It now looked like 06 was enjoying teasing the bears, for she wagged her tail when confronting them.

I noticed how graceful 06 was in her movements. She would get to within a body length of the mother bear, dart off when chased, and come right back when the sow stopped. When the big grizzly tried to smash the wolf with a front paw, 06 would slip away, just slightly out of the bear's range, then return for more teasing. At times the wolf was just inches from the bear, but she always reacted quickly enough to dodge blows and charges. 06 had the same type of athleticism and quick reflexes that Muhammad Ali possessed at the peak of his boxing career. I recalled what one of Ali's frustrated opponents said after failing to land any significant punches: "He was never in one place at the same time!"

By then, wolf and sow were about four hundred yards from the den and pups. After two and a half hours of luring the bears away, 06 seemed satisfied and went back up to the den.

A half hour later, the bears moved back toward the den. 06 howled and must have been trying to get 754 and 755 to come back and help her. She spent another two hours harassing the grizzlies and luring them away, before finally returning to the den area. In total, it took 06 four and a half hours to deal with the threat to her pups. She bedded down in a place where she could spot the bears if they returned and also view the den. I thought that she must have been exhausted after that lengthy ordeal, but she diligently stayed awake and alert in case the bears came back.

When I went over my field notes, I found I recorded fifty-two times when 06 got the bears to chase her. At any moment during the lengthy confrontation, she could have been fatally bitten or knocked out by a blow from a front paw. That was a stunning example of 06's determination to protect her pups. Someone suggested we honor her with the name Fierce Warrior, but we had been calling her the 06 Female for so long we stuck with that.

FOUR DAYS LATER, I saw 06 come out of her den with two tiny pups following her unsteadily. They looked to be about two or three weeks old. 06 tried to push them back into the tunnel with her nose. She got one in, then picked up the other pup and carried it inside. After that, 754 went to the den entrance and sniffed the pups. Then 755 arrived and did the same thing. For both males, that sniffing would likely permanently imprint the individual scent of each pup in their minds. We soon got a count of four gray pups and eventually determined that two were male and two female.

Bear watcher Ralph Neal later saw a lone grizzly come toward the den and all three adult wolves charge in and harass it. I got back there and saw 754 sitting up next to the den, looking like he was guarding the entrance. 06 was bedded down nearby and 755 was charging at the big bear. It countercharged at him. The wolf took off but came right back and bit the bear on a hind leg. Seemingly frustrated, the grizzly left. 06 got up, went to 755, and licked his face as she wagged her tail. It looked like she was rewarding him for his heroic defense of the den. The two males proved they could act as a team to protect the mother wolf and her pups during a crisis.

WHEN I ARRIVED at Slough Creek one morning, I saw 06 below the den site. 755 went to the den entrance and she aggressively charged at him, apparently mistaking him for a wolf from a rival pack. It happened so fast that his scent likely had not registered in her mind, or perhaps her protective instinct overrode everything else. The male ran off, obviously intimidated by the mother wolf. Then both stopped and looked at each other. 06 must have recognized him, for she relaxed and walked over to greet him. That incident demonstrated the effect she had on other wolves, even adult males, when she chose to act aggressively. She was not a female to mess with.

06 continued to outsmart the two young males. I saw 754 carrying an elk leg toward the den. He must have hidden it, for he no longer had it as he approached the site. 06 saw him and got a regurgitation from him. He then went up to the den. The four pups came out and he sniffed them. 06 joined them and nursed the pups. Later she walked off and followed 754's scent trail back along the route he had taken to the den. 06 found the elk leg he had hidden, carried it off, dug a hole, and buried it. The incident indicated to me that no wolf would ever outsmart 06.

By that time, the pups were four weeks old. They were clumsily exploring around the den entrance and playing together. When a pup wandered too far, 06 would pick it up in her mouth and bring it back to the den. 754 must have watched her doing this, for soon I saw him carrying straying pups back to the tunnel. 755 used a different technique. When a pup got too far away, he would get behind it and push the pup toward the den with his nose. Both young males were eager to help out with parenting duties.

Bears continued to come into the den area, both grizzlies and black bears. One day, the pack had to deal with three different black bears. The Lamar adults chased one of them up a tree, and it then fell to the ground. The wolves surrounded the stunned bear and chased it away.

On another day, 06 confronted a black bear on her own. She chased the bear and it chased her. When it moved off, 06 bit the bear on the rear end. 754 ran over to help her. Reaching a tree, the bear climbed up. 06 ran to the trunk, jumped up, and bit it once more on the bottom. She did it again when the bear started to climb down. I wondered how many of the local bears had bite marks from 06 on their rear ends.

By late May, the pups were playing together more frequently and in more varied ways. Wrestling was the most common game, with chasing a close second. I saw one pup sneak up on a sibling and pull its tail. All four looked healthy and alert. For a first-time mother, 06 was doing an excellent job of raising her pups.

EARLY ONE MORNING, I spotted 06 coming back to the den from a hunting trip. She was carrying a dark animal with short legs. When I put my spotting scope on it, I saw that it was a beaver. Wolves in Yellowstone don't often catch and eat beaver because the dam builders usually stay close enough to water that they can dive and escape a predatory wolf. Once in the water, they are fast swimmers and can stay underwater for lengthy periods. Perhaps 06 had surprised this particular beaver on land and caught it before it had a chance to reach water.

On arriving at the den entrance, 06 looked inside and then put the beaver in the tunnel for the pups to eat. But they ran out and nursed on her instead. They had begun to eat meat but seemed to prefer her milk. I later thought her placement of the beaver inside the den was another sign of her intelligence and foresight, for it put the fat animal out of sight of the two adult males—who would have considered it a delicacy.

The big event in Yellowstone that spring was sightings of a grizzly known as Quad Mom, because she had given birth to four cubs, a rare occurrence. I heard that one cub was much smaller than the other three and had a hard time keeping up and dealing with obstacles on their travel route. People saw it climb up on the mother's back when the going got tough and ride along in that position. One time, it could not climb over a large log, and one of its bigger littermates picked the runt up and placed it on top of the log. The following spring, the mother bear came out of her den with three surviving cubs. One of them was the little runt.

ONE DAY 06 had a confrontation not with a bear, but with a bird. I watched as she approached a golden eagle standing on the ground. It had both wings spread over an elk calf it had apparently just killed. This behavior is known as mantling, which eagles display in an attempt to protect their kill from other predators. Undeterred, 06 charged and the eagle flew away. The wolf grabbed the calf and carried it off, but the mother elk was still nearby. She chased the wolf and made her drop the calf. When the wolf later came back, the mother elk was still protecting her dead calf, and she drove 06 off once more. For the rest of day, the cow guarded her calf. By the following morning, the mother elk was gone, as was the calf. She must have realized it was dead and finally left it, allowing one of the wolves to carry it off to the den.

06 CONTINUED TO nurse her pups through June 5. After that, she repeatedly jumped away when pups tried to suckle, a sign her milk had dried up. She and the two males fed the pups by regurgitating meat or carrying parts of carcasses to them. One day I saw a pup approach 755 as he chewed on a bone. The male growled, but the pup ignored him and chewed on the far end of the bone. 755 picked up the bone and walked off with it.

The four pups now were exploring a little farther from the den. One evening I spotted all three Lamar adults protectively following the pups as they wandered around. The adults were wagging their tails, and the pups were active and playful. The little wolves wrestled, had tug-of-war battles, and climbed up and down rock outcrops. When the adults howled, the pups joined in.

I thought a lot about play and concluded that a pup who was good at getting other littermates to play with it was developing social skills that could result in friendships and alliances, especially if the playing was done in a fair and balanced manner. A pup who frequently started play sessions would also be learning leadership skills. 06's great-aunt, 42, had been especially good at making allies and being a leader and 42 likely began developing those skills in play sessions when she was a pup.

ON JUNE 12, I made it to the ten-year mark of going out early every morning to study wolves. Including leap years, that added up to 3,653 days in a row. I looked back on my records and found that I had seen wolves on 99 percent of those days. I averaged about eight hours in the field every day for that period, so that added up to over 29,200 hours. There was a theory going around at that time that a person needed to put in 10,000 hours of training and practice to become proficient at some skill, such as playing the piano. I was close to tripling that amount of time.

Eight days later, Wolf Project researcher Erin Albers got a mortality signal from 480's collar. It came from the Hell-roaring Creek area. I had not seen the former Druid alpha male for a long time and expected we would soon get such a report. A Wolf Project crew hiked out and found his remains in a cave-like opening in a big rock outcrop. He probably was in poor health and found some protection from the weather there. 480 must have died some time previously, for thick grass was growing through his bones. I would like to think he passed away peacefully in his sleep.

In late June, I watched one of 06's pups carry off a piece of meat and bury it for later. Five minutes went by, and then another pup went to the cache and walked away with the buried meat in its mouth, the same type of trick 06 had pulled on the two adult males. After that, I saw two pups chasing another pup who was holding a strip of meat. When the two stopped, the one with the meat ran back and the chase resumed. That showed that it was a game where chasing was the point, not retaining the meat.

During the time I was writing this book, new research on how dogs age was published by a team of scientists, with Tina Wang of the University of California San Diego School of Medicine as the lead author. The study, based on Labrador retrievers, found that young dogs age faster than previously thought. It shows that a year-old dog is comparable to a thirty-year-old person. After that first year, the aging process in a dog slows down. A four-year-old dog would be similar to a person at fifty-two. By the time it is twelve years old, it would be equivalent to a seventy-year-old person. 06 was four when she had her first litter of pups, so based on this new research she was about the same age as a fifty-two-year-old woman.

Yellowstone wolf biologist Dan Stahler told me that female wolves continue to have pups regardless of their age, meaning they do not have the equivalent of menopause. While researching a paper he wrote on the subject, Dan found a record of a wolf in captivity who had pups at age thirteen, and famed wolf biologist Dave Mech has documented a wild wolf who had pups at that same age.

06'S PUPS CONTINUED to explore farther and farther from home, a natural part of their development. In mid-July one of them wandered off by itself and ended up near five large bighorn rams. They looked at the pup and the little wolf ran away, intimidated by their size. Since wolves were brought back to Yellowstone, I have seen them make only one bighorn kill, and it was a young inexperienced ewe. The sheep mainly stay in high, steep country close to cliffs, terrain mostly inaccessible to wolves. That ewe made the mistake of crossing open country and the wolves outran her before she could reach a cliff.

In late July, I got back to Slough Creek and heard that 06 had killed two elk by herself, a cow and a calf. I saw her feeding on one of the carcasses. In each case, 06 leaped up and made the killing bite on the throats of the elk. The cow lifted the wolf off the ground when o6 grabbed her neck but could not shake her off. I had seen other wolves use that same technique, but they were mostly big alpha males who were especially strong. We now knew that o6 could do anything that male wolves could do.

I was beginning to notice differences between the behaviors of the two adult males in the pack. 754 spent more time with the pups than alpha male 755. He often followed them around, and I got the impression that he enjoyed playing with them. The pups seemed to treat him like a giant live teddy bear. 754 was bigger than his brother, so it might seem that he should be the alpha male, but he never challenged 755 and appeared to be comfortable with being the second-ranking male. Later in the season, when the pack began to travel longer distances, I noticed that 754 paid special attention to the pups and seemed to monitor them more closely than 755 did. If the pack was traveling together with o6 or 755 leading, 754 would often position himself last in line behind the four pups, like he was making sure they were keeping up and not wandering off.

BY EARLY AUGUST, the Lamar pups often traveled with the adults. o6 seemed more relaxed now that the pups had reached an age where they could do that. She often played with them and with the two adult males. I saw o6 and 755 playing a chasing game together but she was too fast for him to catch. She sped back and ran circles around him. I looked over at 754 and the big male was romping among the pups, acting like he was one of them. Every wolf in the family seemed happy and joyful.

The Lamar pups had often seen their mother charge at grizzly bears and harass them into leaving the den area. The adults were absent one day when a grizzly came toward the pups. Imitating what they had seen 06 do, all four pups ran at the bear with their tails raised. It saw them and did not seem impressed by the sight of the miniature wolves. Seeing that the bear was not running off, as it would have had the adult wolves charged at it, the pups slowed down and followed the bear in single file, looking like cubs marching behind their mother. I lost them all going through a pass.

On September 21, the Lamar pups scavenged on the elk calf 06 had killed on July 25, fifty-seven days earlier. The next day, the pups hunted for insects and rodents near the carcass. The three adults were not around on either of those days, so the pups had to fend for themselves. We found out that the adults had killed a cow elk in Round Prairie, about fourteen miles to the east. 06 was the first adult to return to the pups. We did not see her arrive, but she likely had gulped down a big load of elk meat and regurgitated it to her pups.

I later saw her go back to Round Prairie, but by that time the elk had been mostly consumed by the other wolves. On reaching the carcass, she fed on the rumen lining and spit out bits of semidigested vegetation clinging to the inside. 06 then ate part of the cow's hide and spat out the fur, trying to get all the nutrition she could from the carcass. Fall can be a tough time for wolves, because most adult elk and bison, as well as their calves, are strong and healthy and difficult to catch after a spring and summer of feeding.

As we entered fall, I felt 06 had proven to be an exceptionally good mother. This was her first litter, but she had helped her own mother raise pups in the Agate pack, so she was experienced and well trained. 06's mother, wolf 472, had been trained in raising pups by famed Druid alpha female 42, who had learned from her mother. That is how it works in wolf families, going back for thousands of generations of daughters apprenticing under their mothers.

06 Moves
to Lamar Valley

L ATE IN SEPTEMBER, 06 brought her pups to Lamar
Valley, the central part of the original Druid territory.
I got signals from the collars on 754 and 755 that indicated the pack passed through the forested area where 21 and
42 had raised many generations of pups.

On October 2, 06 and 755 were on the south side of
Lamar Valley. She was plucking fur from a small carcass and
feeding on it. I took a closer look with my spotting scope and
saw that it was a grizzly cub. 755 was bedded nearby and not
eating. I got reports from wolf watchers that they had seen
the two wolves surround the bear. Then 06 had attacked and
killed it. They had not spotted the cub's mother anywhere in
the area. She might have died or the two might have gotten
separated. 06 did all the feeding and 755 just watched her.
When 754 arrived and moved toward the cub, 755 blocked

him. Then 06 lunged at 755, who had gotten close to the carcass, and gave him a holding bite. After that, neither male tried to interrupt her feeding. It looked like the two big males were afraid of incurring the wrath of a female who had just killed a grizzly. If 06 were a character in *Game of Thrones*, she would be known as Grizzly Slayer.

Without letting either male feed, 06 carried the remains of the bear across the road. We lost her heading toward her pups, who were waiting for her up higher on the slope. Soon after that, I saw one of the pups carrying around the remains of the cub. He put it down and fed on it. Eventually I saw all four pups eating parts of the carcass. I never saw the adult males feeding on it, only 06 and her pups. That incident was another confirmation in my mind of how alpha females are the bosses of wolf families. 06 had killed the cub and she was going to determine who had access to it.

This was a pattern of behavior I saw repeated the following year when her first litter of pups were yearlings. 06 killed a large elk calf with the help of one of them. Another yearling and three of the pups born that year joined them and fed. 754 and 755 were nearby but had not been involved in the kill. The two big males waited seventy-one minutes as 06 and the younger wolves fed, then got up, went to the carcass, and began eating alongside the others. 755's lengthy delay in feeding was one of many incidents I witnessed over the years that proved false the old belief that alpha male wolves eat first and only let other pack members feed once they are full.

06 WAS FREQUENTLY seen up at the den forest that fall. There was a long history of her ancestors, the Druid wolves,

denning at the site, and we hoped 06 would continue that tradition and have her second litter of pups there the following spring.

In mid-October, I heard that 754 and 755 were going downhill from the den forest toward a spot in the road where the Druid pack often used to cross. We called it 21's Crossing. I went to see if the wolves needed help. A car was speeding down the road and I waved at the driver to slow down. As he braked, 755 darted out and the car just barely avoided running over him. The road was a constant obstacle between the den forest and the hunting areas that the wolves frequented to the south. If 06 chose to have her pups in the old Druid den forest, I was going to have to spend a lot of my time being a crossing guard, holding up a big red Stop sign as the wolf family moved across the road.

I saw all the Lamar wolves near my cabin in Silver Gate early on the morning of October 20. They did a lot of howling. When 06 was close to the road, two cars stopped and the passengers took pictures of her. She ignored them. That was a bad sign. Wild wolves have a natural wariness of people, and she seemed to be losing that. I was not worried that she would ever harm a person. The issue was legal hunting of wolves in nearby locations. If 06 did not regain that wariness, she would be an easy target for a hunter.

By that time, 06's family was using Lamar Valley and nearby areas exactly like the Druid pack had for so many years, including the den forest north of the road and the Chalcedony Creek rendezvous site to the south.

I spotted 06 and her pups above 21's Crossing early one morning. She led them down to the road, trotted over the pavement, and continued south. All four pups and both males followed her, just as the younger Druids used to follow 21.

In mid-November, I saw 06 chasing an elk cow. Even though the cow was several times her size, 06 pulled her down. By this time, the pups were old enough to participate in hunts, and two of her seven-month-old pups ran in and helped her finished the elk off. That was proof of what a good job 06 had been doing of teaching her pups how to hunt, as well as an indication that they were fast learners.

The Lamar wolves now ranged as far west as Hellroaring Creek and east to Silver Gate, a territory that stretched nearly forty miles. That was about the same domain the Druids had roamed at the height of their power. But the Druid pack at the time had been much larger. At their peak, they had thirty-eight members. 06 was patrolling the same huge region with just two other adults and four pups.

A FEW GRIZZLIES were still out in late fall. My neighbor Anne Whitbeck told me she saw a grizzly on my front steps early one morning. If I had gone out my door, the bear would have been surprised and might have attacked me or chased me into my cabin. Other wildlife frequented our little town of Silver Gate. I had a moose and her calf and many bison walk by my front window.

Later in November, I was trying to find the Lamar wolves near Slough Creek. The signals from the two males were loud, but it was hard to figure out what direction they were coming from. I had pulled over in a lot alongside the road and was about to get out my spotting scope to look for them.

I happened to turn my head and in the passenger-side mirror saw a gray wolf walking down the road. It was 06. To avoid getting near my car, she veered off the road. Then the four pups and both black males came through and followed her route. Her actions to move away from my car were good to see, especially because this was probably a lesson her pups would remember.

The snow was deep throughout the park by that time. The elk had to expend a lot of energy digging through it with their front hooves to get at the dried grass underneath. Bison use a different technique. They have huge heads and swing them back and forth to clear a crater to feed in. Each swing moves aside far more snow than an elk can with one pass of its hooves, so bison are much more efficient at dealing with snow.

The Lamar wolves were also coping with the heavy snow cover. 755 led through the snow one day, and 06 and 754 followed in the trail he had broken. The pups were on a different route and having trouble. The two lead adults disappeared over a ridge. 754 stopped and waited for the pups to catch up. They reached the trail made by the adults and from then on had a much easier time traveling. As they approached 754, he continued on the route of the alphas, and the pups followed him.

When the pups caught up with the other adults, one of them initiated a chasing game with 06. The pup ran off and 06 chased and pinned it. She playfully gave it a holding bite as it pawed up at her, then let it go. The pup romped off and a second chase began. It outran her, so to continue the game, the pup ran back. The mother wolf caught the pup and wrestled it to the ground. Later, 06 had a similar play session with the rest of the pups. The deep snow seemed to create high spirits in the wolves and make them playful. They looked like a group of neighborhood kids on a day off from school due to a snowstorm.

In early December, the seven Lamar wolves were back at Slough Creek, on a high knoll. Nine Agates were on a ridge to the south and both packs howled at each other.

I always assumed that wolves could identify another wolf they knew well by the sound of its howl. John and Mary Theberge, two Canadian wolf biologists, were doing research in the park on wolf howling. When a wolf howls, the vocal cords vibrate to produce a fundamental tone and numerous mathematically related harmonic overtones. The Theberges told me it is the harmonic overtones of a howl that enable a wolf to identify other wolves they know. To put it in different words, each howl has a main frequency and then bands of sound that deviate from the main, giving it its distinctive pattern. That would be similar to someone listening to a Beatles song and knowing from the sound of the singer's voice if it was John, Paul, George, or Ringo.

472, the Agate alpha female, was 06's mother, so the two of them likely recognized each other's vocalizations. The two Lamar males and the four pups turned around and left the area, but 06 stayed in place longer, looking in the direction of the Agate howls. Then she, too, turned and followed the others out of sight.

I looked at the Agates and saw 472 affectionately licking a pup on its side and neck. The pup was in a relaxed position with its head down, like a dog that was being petted. When

472 stopped, the pup pawed at her to get her to resume the licking. I took that to mean it was pleasurable to the pup. She obliged and started the licking once more. That was the last time I saw 472. She died soon after at the age of nine and a half.

While I was reading the recent book *Yellowstone Wolves*, edited by Doug Smith, Dan Stahler, and Dan MacNulty, I came across an intriguing statement about the sons and daughters of Agate alphas 113 and 472: "The Agate Creek pack, in particular, seemed to have an especially high number of offspring go on to lead their own packs and had inordinate success during inter-pack encounters with strangers." I wondered whether 472's devoted care of her offspring was a major factor in their later successful careers. Her most famous daughter, 06, was certainly proving to be just as dedicated a mother.

754 continued to give 06's pups a lot of attention. He had a tug-of-war with one of them and dragged the pup around on the snow as it stubbornly held on to a tidbit from a carcass. Later he saw two pups having their own tug-of-war. 754 ran over and they had a two-against-one contest, with him on one end and the two pups on the other. He could have easily pulled the item away from the pups but let them win the game.

IN MID-DECEMBER, I got loud signals from the radio collar on the Agate pack's old alpha male, wolf 586, who had taken over that position after the death of 113. Then I heard him howling. He seemed to be up on Specimen Ridge. The Lamars had a big group howl in response. We spotted 586 on the skyline and saw him howling. He then looked toward the sound of the Lamar howls. Perhaps because of old age and a decline in his hearing ability, he mistook the howls as being from his own family and started to move downhill.

Soon he was on the floor of the valley, west of the Lamar wolves. At that time, the Lamars were in the trees. They came out of the forest, saw the old male, and ran toward him. 586 stopped, looked at the pack, then moved their way, apparently still thinking this was the Agate pack. His family had five gray and two black wolves, other than him, and the wolves coming toward him had that same count and color. The Lamars stopped, then approached him in stalking postures.

At that moment, 586 realized his mistake. He turned around and ran off to the north. The Lamars were much faster than the old wolf. They caught and pinned him. At first, he went submissive and did not react when they bit at him. But he soon changed his strategy and fought with them. The fight alternated between him jumping up and battling them and the other wolves pulling him down. 586 broke away and ran off toward the river but had to turn around and face the Lamars when they caught up with him. Despite being outnumbered seven to one, he was doing a good job of defending himself.

586 ran north again with 755 right behind him. The Lamar alpha male bit him on the rear end several times and each time let him go when 586 turned to fight back. When 586 ran off once again, all the Lamars chased him but did not attack.

As the old wolf approached the river, 755 stopped running and his example spread among the other pack members.

They all joined 755 and let 586 go. He went west, then veered off to the south, back toward Specimen Ridge. Soon he bedded down and looked back at the other pack. I saw him eating some snow. He licked his wounds as he watched the Lamars.

I realized that 586's radio collar had given him some protection when the Lamar wolves were biting him, just as 571's collar had helped her when she was attacked. His thick winter coat and tough hide would also have worked to shield him. But more importantly, 755 allowed him to get away and live another day. 755's actions reminded me of 21, and how he had never killed a defeated rival wolf. 06 never met her grandfather, but she had picked an alpha male who had a personality similar to 21's.

The chase and attacks had started at 9:00 a.m. and the Lamars let him go around 9:13 a.m. 586 bedded down soon after that and rested through 5:00 p.m. I then saw him walking off to the south. Then it got dark and I had to head in.

I went there early the next morning and picked up a signal from 586's collar, but there were no signals from the Lamars, meaning they had left the scene of the attack. I spotted 586 in late morning. He was going into the trees on the lower part of Specimen Ridge, not far from where he had been attacked the previous morning. We also got signals from Agate male 641 in that direction, so 586 was heading back to his family.

A few days after his battle with the Lamar wolves, I got 586's signal from the same place on Specimen Ridge where the other collared Agate wolves were. Later that day, the Agates were seen running down a ridge there. 586 was in that group and looked all right. He was a tough wolf who could take a beating and survive.

Several days later, we saw the Blacktail wolves, the pack started by former Druid 302 along with several young Druid males and some Agate females in late 2009. Some of the pups had mange, and a gray male pup seemed to have the worst case. He had lost a lot of fur on his tail. There was a tuft of longer fur at the tip, and so he became known as Puff. Big Brown, a former Druid, and 693, sister to 06, were the current Blacktail alpha pair.

There is one more story to tell from 2010. Shauna Baron, a longtime naturalist and educator in the Yellowstone area, was in Lamar Valley one fall day. She saw a black wolf with a good coat traveling in the east end of the valley with a gray female. On the chest of the black was a blaze in the shape of a triangle. Was it the mangy but heroic young Druid wolf who had risked his life to save his sister, 571? We do not know for sure, but I like to think it was.

The Risk-Takers

I have great admiration for risk-takers, whether human or wolf. Most young female wolves stay home in the safety of their family's territory. Some might eventually pair off with young males who come into the territory. It is the especially courageous daughters, such as 06, who choose to leave the security of the family's homeland and head out into unknown terrain as lone wolves. They are the adventurers of their species.

The first one to do this in Yellowstone was wolf 7, a pup who came down from Alberta with her mother, Rose Creek alpha female 9, in early 1995. Soon after being released from the acclimation pen in March, 7 left her mother and her mother's new mate, male wolf 10, and took off on her own. At that time, she was about twelve months old. As far as I know, she was the youngest of our females ever to do that.

She lived as a lone wolf from that spring through January of 1996, when she ran into wolf 2, a yearling from the Crystal Creek pack, and paired off with him. The new pair established a territory west of Tower Junction. They were named the Leopold pack in honor of Aldo Leopold and his 1944 suggestion to restore wolves to Yellowstone. In April 1996, wolf 7 had her first litter of pups. She was two years old at that time.

As she had left her family when she was so young, wolf 7 had no experience in helping her mother raise younger siblings, the normal process that prepares females for raising their own pups. She had to figure everything out by herself, and she did a masterful job. Her first litter in 1996 had three pups. I spent a lot of time that year watching the new mother and her pups. It was the first time I was able to study a mother wolf in such detail.

I saw that 7 was very interactive with her young pups and often played with them, especially in games of chasing and wrestling. If a pup tired, she would try to keep the fun going by jumping up and down in front of it and doing play bows. As she was still young, she often played by herself. I saw her repeatedly throw a piece of meat into the air and catch it on the way down. At times she would run in circles for no apparent reason other than it must have been fun.

She also played with her mate, wolf 2, and they were especially affectionate to each other. I saw her stroke his face with a front paw, then lick his face. Despite how young she was, 7 was a perfect mother and mate.

She eventually gave birth to seven litters and had a total of thirty-nine pups. Wolf Project records show that at least twenty-nine of those pups survived to their second year, a tremendously successful record considering how inexperienced she was when she first became a mother. One of her sons was the famous wolf 302. Years after his mother's death, he returned to the family's territory and started his own pack there, the Blacktails. The pups he sired and raised in the old Leopold territory were the grandchildren of wolf 7.

Since the days when 7 was the first young female in Yellowstone's population of reintroduced wolves to leave the security of her family, other young Yellowstone females have done the same thing. In the early 2000s, when the Druids were a superpack with thirty-eight members, I kept track of the adventurous daughters who, like wolf 7, chose to risk leaving home. Some failed and soon returned. Others were never seen again and likely died. Only a few succeeded.

Druid female 217, a daughter born in 2000 to 21 and 42, was one of the successful ones. She left her family in late 2002 when male wolf 261 from the

Mollie's pack came into her territory. She paired off with him and they formed the Slough Creek pack just west of her family's territory.

Wolf 472 was also born in 2000. Her parents were 21 and 40, but since 40 died during the first few weeks of that denning season, 472, like 217, was raised by 42. I always felt that 472 would have regarded 42 as her mother, since she likely had no memories of 40. In early 2002, 472 helped form the Agate Creek pack with some of her Druid sisters and eventually became its alpha female.

472 lived a long life and raised many litters of pups. Several of her daughters were risk-takers like herself and became famous alpha females, including the 06 Female and wolf 693, who cofounded the Blacktail pack. A third daughter, three-and-a-half-year-old 715, took over the Agate alpha female position after the death of her mother. One way to evaluate the life of a mother wolf is to look at the accomplishments of her daughters. By that standard, 472 was supremely successful.

PART II

2011

6

06's Second Litter

A T THE START of 2011, the seven wolves in 06's family were spending most of their time in Lamar Valley but often traveling west a few miles to Slough Creek. The neighboring Agate pack had eight wolves and there were fifteen in the Blacktail group, who were based west of the Agates. The wolf population of Yellowstone at that time was estimated to be ninety-six, with eleven packs and a few loners. There were fifty-nine adults and thirty-seven pups in that population. Over the years, there has tended to be an average of about one hundred wolves in the 2.2 million acres of Yellowstone National Park.

Doug Smith, the lead biologist in Yellowstone's Wolf Project, did some radio collaring in early January. Those operations were done from a helicopter, where Doug would shoot a wolf with a tranquilizer dart. That day he put new collars on Agate males 586 and 641. Both were big and weighed in at 120 pounds. They originally were in the Mollie's pack, so they were descended from Crystal Creek wolf 5. Wolf 586 had some broken and infected teeth, a common occurrence for an adult wolf. The crew then went east and darted a gray Lamar pup. She was 70 pounds and would be known as wolf 776.

The Wolf Project tries to have at least two wolves in each pack collared, so we can keep track of their whereabouts. When the project biologists collar the wolves, they also weigh them, check their health, and take genetic samples so their ancestry can be charted. Each collared wolf is assigned a number for record keeping. Uncollared wolves are not numbered. Instead, we refer to them either by an easily identifiable physical feature—as with the wolf known as Big Blaze—or, as in the case of the 06 Female, by some biographical aspect that sets them apart from other wolves. Radio collars are sometimes chewed off by fellow pack members and batteries last for just a few years, which means some wolves are collared more than once over the course of their lives. The collars send out signals we pick up with a handheld antenna. The signals can be detected from several miles away, but the topography of the area can limit how far the signals travel. I spent a lot of my time scanning areas with my antenna to see what signals I could pick up and then trying to visually locate the wolves I was hearing.

IN MID-JANUARY, THE Agate and Lamar wolves howled at each other from opposite sides of the valley. Soon both packs moved off, away from each other. Based on that sighting and the earlier incident when the Lamar wolves spared the life of Agate alpha male 586, I concluded the two packs had some sort of understanding about tolerating each other.

A few days later, an Agate black pup got separated from his family and ended up near the Lamar wolves. The pup did a lot of howling. The Lamar pack howled back but seemed unconcerned about having him close by. They could have easily caught and killed the pup but left him alone. That was another indication of what seemed to be a truce between the neighboring packs, a truce that went all the way back to the time when 21 and 113 were the respective alpha males of the Druids and Agates. The Agates included alpha female 715 and her older sister 471, both of whom were sisters of 06. 471 had been in the Lava Creek pack with 06 before that group broke up. 06 had grown up in the Agate pack, so the two alpha females would have good reason to continue that tradition of toleration.

A related issue was that the Agate territory was positioned between the Lamar and Mollie's territories. That meant the Agates served as a buffer that might hinder the Mollie's from reaching Lamar Valley, their original territory.

In late January, Doug collared Blacktail alpha male Big Brown and assigned the number 778 to him. He was another big wolf: 119 pounds. Like the other adult males in the Blacktail pack, he was born a Druid.

My job during collaring operations was to help the helicopter crews find the wolves and explain to park visitors how our research and monitoring program worked. The people I spoke with were always extremely excited to see the helicopter darting and collaring take place. It was like watching an action movie take place in front of you.

A few days later, the crew collared a huge male in the Delta pack. He turned out to be the biggest wolf ever weighed in the park: 147 pounds. The average weight of adult wolves in Yellowstone is around 100 pounds, with males being bigger than females. That Delta wolf was an outlier and just happened to be way past the average. There is a lot of variation in the size and weight of wolves, just as there is with people. For example, multitime WWE heavyweight champion Drew McIntyre is six feet, five inches tall and weighs 265 pounds, while I am five feet eleven and around 145. Drew and I are both 100 percent Scottish and therefore have somewhat similar DNA; he was just destined to be much bigger than me. So far, no one has mistaken me for him.

There is a disadvantage if a wolf is too heavy. Wolves are pursuit predators. They need to chase and catch prey animals like elk that are normally faster than them. But bison are the main prey of the Mollie's wolves in winter, and the huge animals, which range up to 2,000 pounds, often stand their ground when wolves approach. For that reason, it was an advantage to have bigger-than-average males in the family. The founding alpha female of that pack, wolf 5, must have had genetics that tended to produce big male descendants.

WE HAD A cold snap in early February. The low one morning at the Lamar Ranger Station was minus 37 Fahrenheit (–38 Celsius). The following morning, it was minus 42 (–41 Celsius). The wolves I watched that day were bedded down and, thanks to their thick, mange-free coats, seemed oblivious to the extreme cold. I had on many thick layers but was freezing.

755 and 06 mated on February 10. A few days later, 754 went to 06 and mounted her. 755 saw what was happening and knocked his brother off her. Later that day, 755 bred

with 06 again. On the last day of February, the Lamars were once again at the den forest. 06 was likely pregnant at that time, and I guessed that she had decided to den there.

An alpha female's choice of a denning site is likely the most important decision of the year for a wolf pack. The right choice will mean a high pup survival rate, while a wrong choice could result in loss of the entire litter, a terrible disaster for a wolf family. The previous year, 06 had denned at Slough Creek and had 100 percent survival of her pups. I did not know why, but apparently she now felt she had found an even better place to have the next set of pups.

Of the two brothers, 754 was the one that spent the most time playing with the pups, but in early March I saw 755 in a play session with a pup from 06's 2010 litter. 755 ended up on the ground with the pup standing over him, looking like he had just defeated the big male in battle. Then 755 jumped up and romped around, acting like a young pup himself. He then went to another pup and did a play bow to it, an invitation to play together.

A few days later, when 754 was playing with the pups, 755 ran over and joined in. A pup nipped him, and the male chased the little wolf. Then that pup turned around and sparred with 755. The two acted like equals during the play session. 755 seemed to be pretending he was just one of the pups, rather than the alpha male. As with human families, there is a time for playing and a time for working. Both adult brothers would soon be spending long, exhausting hours trying to feed 06's new litter. But by then last year's pups would be yearlings, and they could assist the two adult males on hunts, as well as in bringing back food for the growing family.

THE WOLF PROJECT'S March Winter Study was going on at that time, a thirty-day research project that had teams of biologists assigned to three different packs. They first used telemetry equipment to find their study pack, then recorded everything the wolf family did. Most of them were volunteers. I joined the crew assigned to the Blacktail pack and saw the wolves feeding on a fresh cow elk carcass. The crew told me the ten-and-a-half-month-old pups did most of the work in getting the cow.

The elk had gotten bogged down in deep snow. The pups had seen her struggling and charged that way. The cow got free of the snow and ran off, then hit another section of soft snow. A female pup ran in and grabbed the cow's throat but lost her grip. A male pup arrived and got a holding bite on a hind leg as his sister bit into the cow's throat again. That pup pulled the cow down by herself. Another female pup ran in and also bit into the throat. Two more pups rushed in. The cow managed to get up, but the five pups jointly pulled her down. Alpha male 778 came in at that point. The pups were about to finish the elk off and probably did not need his help. They had worked as a team to make that kill and the females were the star players in the hunt.

At that time, a high school student was volunteering with the Wolf Project as she worked on her senior thesis. She told me she had followed the tracks of the Lamar wolves through the snow and discovered a site where they had attacked and wounded an elk. Later she found a dead cow elk at the base of a cliff, and the nearby tracks showed that the cow had made it to the top of the cliff, fought with the wolves there, then fallen to her death. I went out there and confirmed everything the student told me. I was impressed by her initiative and tracking skills.

Things were warming up by March 10. That day, it got up to 48 Fahrenheit (9 Celsius), 90 degrees higher than the low we had in February. The first grizzly was seen on March 14, feeding on a bison carcass in Round Prairie.

IN MID-MARCH, I took a close look at 06 with my spotting scope and saw that her sides were sticking out, a sign she was pregnant. During that time, she would often leave the rest of her family and go to the den forest by herself, likely to ready her den for the birth of her pups.

The long winter we had just gone through, with temperatures down to minus 42 Fahrenheit (–41 Celsius), had taken a toll on the prey animals in the park and weakened them. The snow was still deep throughout Lamar Valley. On March 19, I saw a bison bull eating the needles from a Douglas fir tree. Most of the needles on nearby branches had already been consumed. Those fir needles would have little food value to the bison, so his consumption of them was a sign he was not able to find much else of nutritional content. Early in the morning four days later, I found that same bull bedded down on the park road. He did not react to cars passing by him within a few feet. That was another sign of his deteriorating, lethargic condition. The black asphalt absorbs warmth from the sun during the day and retains a bit of it overnight. That slight warmth probably had drawn the sickly bull to the road.

Around that time, I saw 06 heading back east toward her den by herself. On the way, I spotted her approaching a large elk calf that was probably about nine months old. It saw the wolf and moved off, but soon stumbled. 06 must have taken that as a sign of weakness, and she ran at it. The calf tried to countercharge her, but the wolf easily dodged it. 06 leaped up and grabbed its throat, then pulled it down. The calf tried to get up, but the wolf maintained her iron grip and kept it down. At the one-minute mark, the calf began sliding down the snowy slope on its side. 06 moved along with it, always holding on to the throat. It was dead within four minutes of the encounter. That calf must have been in poor health, like the bull bison I had seen eating fir needles and gathering warmth from the park road.

06 was about halfway through her pregnancy, and the fight with the calf must have taken a lot out of her. She walked off, bedded down, and went to sleep. About forty-five minutes later, she got up, went down to the carcass, and started to feed for the first time.

WHILE I WAS writing this section, there was a lot of news in Montana and Wyoming about a debilitating sickness in elk and deer called chronic wasting disease (CWD). It is spread from infected animals to healthy ones and can potentially be transmitted to humans through ingestion of meat. CWD is similar to mad cow disease in that it affects brain function and is incurable. Lately the infection rate has been increasing at an alarming rate. In early 2021, infection in white-tailed deer was found to be 50 percent in one part of Montana.

I do not know if that elk calf had CWD, but it had stumbled as 06 approached, and that unsteadiness is something you would see in an infected elk. Wolves seem to be immune to CWD, so when they kill weakened, infected animals and feed on their carcasses, they help limit the spread of the disease. Dr. Andy Dobson of Princeton University, an expert in diseases that can jump from animals to people, stated in an interview with Todd Wilkinson on the website *Mountain Journal*:

> If you have wolves and coyotes they kill and consume these weakened animals and effectively remove CWD from the ecosystem.... Wolves and coyotes are our strongest defense against CWD, particularly wolves—they are pursuit predators who always focus on the weakest animals in a group of potential prey. As CWD manifests itself by reducing locomotory ability, wolves will key in on this and selectively remove those individuals from the population. These animals are then not available to infect uninfected individuals in the herd.

After reading that interview, I came across a research paper entitled "The Role of Predation in Disease Control," with Margaret Wild as the lead author. She and her colleagues studied how selective predation by wolves on deer that were weakened by CWD infection might greatly reduce transmission of the fatal disease to other deer in the area. Deer and elk infected by CWD are not only weakened but far less vigilant regarding predators. Every infected animal that wolves kill and consume ends the possibility of that animal transmitting the disease to others. The researchers found that wolves have the potential to accomplish an "elimination of the disease" in a given area.

My Park Service colleague in Yellowstone Norm Bishop, who had the critical job of explaining our wolf reintroduction program to the public in the 1990s, later summed up the current CWD issue with this statement: "Wolves are on the hunt 365 days a year. They are our last, best hope to reduce or eliminate chronic wasting disease."

A FEW DAYS after the incident with 06 and the calf, I spotted 754, 755, and the pups on a new carcass. The wolves howled and I heard an answering call from the den forest. 06 arrived eight minutes later. The howling must have alerted 06 that her pack had food, and she went right to them. It seemed clear to me that wolves can convey information through howls, in this case about a new kill. It was similar to how cooks on farms and ranches will go outside and ring a bell to tell everyone that food is being served.

As I watched 06, I saw that she was not acting in her normal dominant manner. Instead, she was behaving like a pup or a low-ranking adult. 06 lowered her head, went to 755, and licked his face. She pestered him for a feeding, even though the fresh carcass was just a few yards away. He regurgitated a big pile of meat for her. I thought that she might be behaving that way to train the other wolves to bring food to her at the den, when she would be caring for newborn pups and unable to leave them to go to distant carcasses.

I often got signals from 754 and 755 from the den forest. That was where Druid alpha female 40 had had pups in 2000, including a female who would grow up to be 472, the alpha female of the Agate pack. It looked like 06 would have her pups in the den where her grandmother had given birth to her mother.

I thought about how 06's mother was born to 40 but was raised by 42, who had a vastly different personality than her sister. 40 was a tyrant to the other females in her family and prone to use violence to get her way. 42 had a cooperative style of leadership. If 40 had a heart of stone, then her sister had a heart of gold. I had known 472 for many years and saw that she took after 42, the wolf who had raised her, not her biological mother. Since 472 was 06's role model for being a mother and alpha female, it seemed that the benevolent influence of 42 extended through 472 to 06.

THE WARMER WEATHER in early April was making the deep, soft snow difficult for the wolves to travel across. At times I saw that the entire length of their legs would sink into the snow. These conditions would exhaust the wolves, as well as the prey animals. Eight inches of new snow fell two days later, making travel even harder for animals. One of my neighbors in Silver Gate told me she measured fifty-six inches of snow on the ground at her cabin. After that, we got another seven inches.

By that time, we needed a system to tell the four gray Lamar pups apart in our field notes and records. 776 had lost her collar but had distinctive markings that enabled us to identify her. The other female became known as Middle Gray. The two males had tones of gray fur that differed in shading. They were known as Light Gray and Dark Gray. They were different wolves than the dispersing males with similar names who had interacted with the Druid wolves in 2007 and 2008. Since the collars on 754 and 755 were still working, we could track the pack using hand antennae.

We spotted a fresh bull elk carcass in the willows south of 06's den early one morning. The four pups were uphill from the road, looking that way. 755 crossed the road and went to the bull elk. Because the carcass was so close to the road, a lot of cars were stopping and 755 did not stay long.

I contacted the law enforcement rangers and two of them responded to my call. We pulled the bull elk farther from the road. I checked his femur bone marrow and found it to be in poor condition. It was reddish and gelatinous rather than whitish and solid, which would have indicated a good fat content. The bull had drawn down the fat reserves in his body, then in his marrow, and was probably close to death when the wolves found him and finished him off. Later, 776 went to where the bull had been before we moved it, followed the scent trail to its new location, and fed—a good example of a wolf's ability to figure things out. She and the other pups born in 2010 were twelve months old, so I will call them yearlings from now on.

06 STAYED IN the den with her newborn pups from the sixteenth to the twenty-fifth of April. On the twenty-sixth she was seen chasing bighorn rams near her den, meaning she had recovered so well that she had resumed hunting. I saw her the next day and noticed that she had a lot of missing fur under her belly and had distended nipples, signs she was nursing.

The Lamar adults were in good spirits in early May. They found a patch of snow and 754 slid downhill on it, then rolled on his back farther down the slope. 755 saw his brother doing that and did the same thing. After that, he ran around in a romping gait, looking like a pup.

Wolf Project biologist Colby Anton and I later examined another bull elk the Lamar wolves had fed on. Using a bone saw, I cut into a leg bone and discovered that the marrow was so depleted that it was liquid rather than the normal solid state. I sawed into other bones from that bull and could not find any marrow at all. Like the bull we had found near the den, he was in such poor condition that if the wolves had not killed him, he likely would have soon died on his own. Both of the bulls were good examples of how wolves are selective predators: they select for individual elk that are in the poorest physical condition. Strong, healthy elk can usually outrun or outfight wolves.

A few days later, 06, 755, and the yearling we called Dark Gray went out on a hunt. Dark Gray found a dead elk calf that may have been stillborn. The yearling carried the calf off and put it down. He must have been full, for he did not feed. I left and Laurie Lyman continued to observe the wolves. She told me that when the young wolf moved off, 755 walked toward the calf. Dark Gray saw that, ran back, growled at the alpha male, grabbed the calf, and carried it away. He fed on it for an hour as 755 watched. When the yearling was full, he stepped away. Only then did 755 come back and scavenge for scraps. That reminded me of the time years earlier when Druid wolf 42 caught an elk calf and her aggressive alpha female sister, wolf 40, did not try to take it from her. I took those incidents to mean that on both sides of wolf hierarchy, male and female, ownership rights to food are respected by high-ranking wolves.

Soon after that, the Lamar wolves killed a bull elk south of the den forest. 755 tore off one of the legs and headed up to the den to give it to 06 and the pups. They would be about six weeks old by then and probably had been eating meat for a week or two. I figured 755 had stuffed himself at the carcass, so he would also regurgitate that meat to the mother and her litter. It was extra work for him to carry the heavy elk leg up there, but conveying the double load demonstrated how dedicated he was to fulfilling his responsibilities to feed his family.

A married couple was visiting the park at that time with their ten-year-old daughter Rio, who was blind. Ken Sinay, who owned a local wildlife guiding company, had taken Rio and her family on a backpacking trip up the Lamar River. One of his other guides would be escorting them up to the Rose Creek acclimation pen the next day. That was the pen where the original members of the Druid pack had been placed when they arrived from British Columbia in early 1996.

I had tremendous admiration for Rio and wanted to do something special to give her a connection with the wolves. Normally I would show people photos of wolf 21. This time, I took out a cast of his paw and handed it to Rio to feel. I then put a sculpture of 21 made by local artist George Bumann into her hands. She touched the cast, felt the wolf sculpture, smiled, then looked in the direction of her parents with an expression of awe. That was one of the best moments of that year for me.

I SAW 06'S pups for the first time in late June. There were five: three blacks and two grays. They ran around and playfully snapped at each other.

In my books I frequently mention the Footbridge parking lot, which is located across the road from the den forest.

Its name comes from a small bridge nearby that spans Soda Butte Creek. From there, a hiking trail goes south up the Lamar River toward Cache Creek and other tributaries. In early July, when Soda Butte Creek was deep and fast-flowing from snowmelt, female yearling 776 came in from the south and approached the bridge. Wolves normally swim the creek to get to the den farther to the north. But this time 776 went to the southern end of the bridge and sniffed it. She immediately went into a crouch and tucked her tail. I thought that the human scent there was making her wary.

Despite that, she took a few steps out on the bridge, but then turned and ran back to solid ground. The wolf seemed to be listening intently, probably for suspicious noises, then she went all the way across the bridge in a crouch. After that, she ran across the road and headed up to the den. I was impressed. Most wolves do not seem to understand the concept of a bridge.

THAT WEEK, A bison died at Slough Creek and the seven adult Lamar wolves regularly traveled the eight miles from their den to the carcass. Elk herds were grazing on the lush grass in the meadows around the creek. One day the pack chased a group of cows. The wolves and elk went out of sight behind a hill, then most of the elk reappeared past the knoll. They bunched up and looked back. The wolves did not reappear, a sign they had gotten something.

About an hour later, I saw 06 coming out from behind the hill with an elk leg in her mouth. She swam through a portion of the flooded Slough Creek and kept her head high as she held the leg above the water. Stopping on a strip of land between two wide channels, 06 dug a hole and buried the leg. After she had cached the leg, she swam across the second channel and headed toward the den to feed the pups with regurgitated elk meat. Her belly was so full it looked like it could not hold another ounce of meat.

Her plan of hiding the leg did not work well. One of her yearlings arrived at the cache a half hour later. She dug out the leg, fed on it there, then swam across the creek holding what remained of the leg in her mouth. On the other side of the water, the wolf dug a new hole and buried the leg by pushing dirt over the hole with her nose. I recalled that when 06 denned near there last year she often stole food buried by /54. It looked like this yearling had learned the same trick and used it on her mother.

A week after we saw 776 cross the Soda Butte bridge, I got a report that 06 had just walked across it to the north side. I then saw male yearling Dark Gray follow his mother's scent trail to the bridge and confidently walk to the far side. Either 06 had learned about the bridge from her daughter or had figured it out on her own, and her son followed her example. A few days later, I saw 06 trot over the bridge, confirming the earlier report. I also once saw 06 pause at the road and then look both ways for cars before crossing to the other side. Not only had she mastered bridges, it seemed she also understood the concept of two-way traffic.

A HUGE BISON died at that time, south of the den forest, and a big grizzly took over the carcass. Grizzlies will often commandeer a carcass for days, sometimes even sleeping on it to keep smaller bears, wolves, and coyotes away from their prize. The wolves and coyotes would often wait for an opportunity to sneak in when the grizzly was not looking or was distracted by something else. I saw 755 approach the dead bison. The wolf bit the grizzly on the rear end, but the bear ignored him, apparently more hungry than annoyed. 755 moved around to the far side of the bison and fed there. The compromise seemed to work for both of them.

Wolves have a great deal of practice in dealing with grizzlies and can outrun them. People are far slower than a charging bear, which can reach speeds of up to forty miles per hour. Soon after watching 755 work out the compromise with the grizzly at the bison carcass, I stopped at a parking lot east of the den forest and saw that two women were walking out toward Soda Butte Creek. They looked back at their young children still in the lot and yelled at them to come out and see grizzlies on a moose carcass.

With bear spray in my hand, I went out to the women, saw two grizzlies on the nearby carcass, identified myself as a Park Service employee, and explained that it is dangerous to approach bears when they are feeding. The women agreed to go back to their families. Four days before that incident, a mother grizzly in another section of Yellowstone had killed a hiker, and the following month a second hiker was fatally mauled by the same mother bear at a carcass. For public safety she had to be euthanized, and her two young cubs were donated to a wildlife park.

Yellowstone is famous for wolves, bison, and grizzlies, but not all of its wildlife is in the megafauna category. One day I was in the Footbridge lot and saw a tiger salamander approaching from the south, the direction of the creek.

Mature tiger salamanders in Yellowstone can grow up to nine inches long, but this one was less than that, perhaps five inches or so. It very slowly moved north, toward the road. I figured it was a male who had just come out of hibernation and wanted to get to a pond up in the den forest, where there were likely female salamanders to breed with. Cars and trucks were regularly driving through that section of the road and there was a high probability the salamander would be run over.

I had the big red Stop sign in my car that I used to manage traffic when the wolves wanted to cross the road, but I could not hold up drivers for the lengthy time it would take the salamander to get to the other side. Then an idea came to me. I got the flat metal sign and set it down right in front of the little amphibian. He got up on it and slowly walked toward the far side. Trying to keep the sign as steady and horizontal as possible, I rushed across the road and put it on the ground. By that time, the salamander was within an inch or two of the far edge, and he calmly stepped off and went on his way. I had previously been asked to be in a reality show about Yellowstone but had turned the television company down. If we had done that series, the salamander rescue could have been part of an episode.

IN LATE JULY, I saw the Lamar wolves hunting elk at Slough Creek. All of them were too fast for the pack to catch. The wolves gave up, went to a nearby four-month-old bison carcass, and scavenged on scraps. The wolves were still hungry the next morning and returned to that carcass to feed. 06 spotted a group of cow elk that had two calves in the herd.

She charged forward and targeted a calf. The calf's mother and other cows countercharged the wolf and ran over her before 06 could get out of their way. Despite being trampled by numerous 300-to-400-pound animals, she jumped right up and resumed chasing elk like nothing had happened. Once again, I was impressed by the incredible willpower and physical resiliency of 06.

BY EARLY AUGUST, we could see that 06's five new pups were thriving and active. They also had big appetites. One morning I saw 06 feeding on a new carcass about six miles west of the den. When she was full, she went to a nearby bush and wiped her face clean, then headed back to her pups. Seventy-one minutes later, I spotted her arriving at the den forest. We saw the pups run to her for a feeding.

She was back at the distant carcass three hours later, fed for twenty-eight minutes, and trotted back to the den to give her pups another feeding. The combined distance of those trips was twenty-four miles. In the evening, she made another round trip to the carcass, making her total distance for the day thirty-six miles. The trampling incident had shown me several aspects of 06's character, and this sighting documented how dedicated she was to keeping her pups well fed.

Later that week, there were no fresh kills, so 06 went back to an old bison carcass and scavenged on what she could get for her pups. A grizzly arrived and chased her off fifteen times. After each chase, the wolf came right back to the carcass. In frustration the grizzly gave up and left. 06 resumed feeding and then headed back to the den. The following morning, I found 754 getting what leftovers he could from another old bison carcass.

When I watched the five pups, I noticed that one of the three blacks tended to be a leader and later wondered if she would grow up to be an alpha female. I also picked up on how often 754 was with the pups. Just as with the first litter, this new group of pups seemed to especially like him and would follow him around. 754 acted like he enjoyed being around the pups and it seemed to be a natural part of his personality. He might not have been cut out to be an alpha male, but he was a good caretaker of pups. Each of the three adults—06 and the two brothers—had their own personal strengths and skills and together they were a superior team.

Mange was an issue for the Yellowstone wolves in those years, and the fact that 755 was mange-free might have been one of the reasons the Druid females had greeted him so enthusiastically and welcomed him into the pack before he decided to leave and join 06. Many of the Druid females had mange back then, and the disease contributed to their pack falling apart. We got good news in August: only one wolf in Yellowstone was known to have mange and it was a mild case.

BY LATE AUGUST, the yearling we called Dark Gray was often away from the Lamar pack. I felt he was probably thinking of leaving to find a mate and start a family of his own. The three remaining yearlings were playful with the pups and engaged them in lots of games of chasing and wrestling, the favorite forms of play for wolves. During one of those play sessions, I saw that 06, 754, and 755 were bedded down nearby and watching the yearlings play with the pups. The three adults were raising their second litter of pups with a lot of help from last year's pups. By any standard, o6's pack was very successful.

Early on August 30, I spotted the Lamar wolves south of the road, up the Lamar River. All five pups were traveling with the adults. Two people on horseback saw the wolves and rode toward them. All the wolves, including the pups, seemed afraid of the horses and people and ran away. In contrast, later that day the pups and adults chased a grizzly. The bear soon stopped, and the pups casually walked around it. The pups had already learned how to handle themselves around grizzlies and did not seem afraid of them. But the horses were something new and their instinct was to avoid them. That was a good thing, for most of the hunting parties outside the park used horses.

7

Prelude to War

WOLF PROJECT RESEARCHER Erin Albers did a tracking flight in early September. Tracking flights are done in fixed-wing planes for the main purpose of checking on the locations of radio-collared wolves. Erin got a mortality signal from the Mollie's pack alpha male, wolf 495, in his family's territory south of Lamar Valley. A Wolf Project crew hiked out to examine his body. Colby Anton, a member of the crew, later told me there were no bites or wounds on him, but there was blunt force trauma to the stomach, likely caused by a kick from a bison. During the necropsy, the crew found signs of hemorrhaging under the fur and a ruptured spleen, more evidence he had been kicked, probably several times. It appeared 495 had lain there for several days before dying, and there were signs that other Mollie's wolves had bedded down near him during his final days. That part of the story reminded me of stories about dogs loyally staying by the side of their dying human friends.

The death of 495 created a problem for 486, the pack's longtime alpha female. As far as we knew, all the other adult males in the family were her sons. Since wolves normally do not breed with close relatives, she would have to recruit a new alpha male in the next few months. There were twelve adults and seven pups in the pack, an intimidating count if a lone male got the scent of 486 and figured out that she was without a mate. Her adult sons might attack any strange male that approached the family. 486 left the pack in late November, probably to seek out an unrelated male. Her three-year-old adult daughter 686 took over the vacated alpha female position, but she too had a problem: all the males in the nineteen-member family were likely her brothers.

The ascension of 686 to the alpha female position set off a chain of events that resulted in a time of great violence and upheaval. The Mollie's were the traditional rivals of the Druid genetic line because of the incident in 1996 when the original Druid wolves attacked the ancestors of the Mollie's wolves in Lamar Valley, killing their alpha male and all the newborn pups. Wolves in the Lamar, Agate, and Blacktail packs were descended from Druid wolves. Over years since the original attack, the Mollie's wolves regularly came back to Lamar, their ancestral homeland, and had rematches with the Druids and their descendants. They killed Druid alpha female wolf 42 during one of those trips in early 2004. If the Mollie's returned to Lamar Valley, the pack would present the most dangerous threat from wolves that 06 would ever have to face.

The Agate pack came into the Chalcedony Creek rendezvous site later in September. The Lamar wolves chased them away and there was no fighting between the packs. Once again, the truce between them held, probably because members of the two packs were closely related. During the incident, the Lamar pups went up to the den across the road to the north, a good instinct when a rival pack was in the area. Later 06 and 755 visited the den to check on the pups.

WE SOMETIMES HAD to put up orange traffic cones along sections of the road. The wolves were fascinated by the rubber objects and would often grab one and walk off with it. I saw a Lamar yearling pick up a cone in its mouth, carry it around, then place it upright back on the road, but in the wrong place for our purposes. Soon after that, I saw a black pup carrying a cone up in the den forest. A yearling had probably brought it up there and given the cone to the pups to play with.

After the pups were old enough to travel and the pack left the den area, a ranger walked up to retrieve that cone. In the meadow, he also found eleven elk antlers the adult wolves must have carried there and given to the pups. All the antlers had small bite marks on them, so they, along with the cone, had served as chew toys.

I noticed that when the entire Lamar pack was traveling together, 754 was frequently last in line, behind the five pups. Once again, I got the impression that he deliberately positioned himself there to watch over the pups and make sure they did not stray off the travel route. He was acting somewhat like a border collie herding a flock of sheep. I also saw that 754 frequently bedded down near the pups and watched them play. The pups were drawn to 754 and would often go over and greet him. I saw his tail thumping the ground repeatedly as a pup licked him in the face, a sign he enjoyed interacting with them.

Another bison bull died south of the den forest. He was near a hiking trail, so the rangers had to drag him farther away, as grizzlies would likely soon arrive to feed on the carcass. Getting the bison away from the trail also enabled the wolves to feed without being near people. After feeding on the carcass in its new location, the pups gravitated to a nearby rendezvous site the Druids had previously used, on the lower west side of Mount Norris. The pups seemed to prefer that to the Chalcedony site, so the adults went along with where the pups wanted to be.

IN MID-SEPTEMBER, I saw the pups join adults on a hunt for the first time. Three adults and three of the pups chased a big bull elk, but he easily outran the wolves. Then I spotted a pup chasing a pronghorn antelope. Pronghorns are the fastest animals in Yellowstone and can accelerate to sixty-five miles per hour. Most wolves have a top speed of thirty-five miles per hour, so the pup was left in the dust. But that young wolf may have learned something: pronghorns are too fast to catch.

After that, the three pups saw a big bison bull and trotted toward the massive one-ton animal. A nearby yearling watched the pups, probably out of concern that they might be trampled. The pups wisely ran off when the huge bull came toward them. But two came back, circled the bison, and approached his rear end, something an experienced adult wolf might do. Then, realizing the big bull was too much for them, the pups wandered off.

As I thought about that sighting, I realized that interactions like this would have played out countless times in past centuries when vast herds of bison ranged throughout the West, times when wolves were the dominant predators in the same regions. My first published book was titled *Denali National Park: An Island in Time*, and I came to see that subtitle also applied to my years in Lamar Valley. When I was out there watching wolves, bison, elk, pronghorns, and grizzlies, I was living on an island in time. Yellowstone today has the largest free-ranging bison herd in North America, and modern scenes of wolf families hunting bison are exactly like moments thousands of years ago in the valley.

I sat in on a class Doug Smith was teaching at the Yellowstone Institute, and I was especially interested in what he said about the digestive system of bison. They have evolved a longer gut than elk. That enables them to hold plant food longer in their system and digest it more efficiently than elk do. In the spring, grasses in places like Lamar Valley reach their peak of nutrition about five or six weeks into the growing season. After that, they lose much of their food value to grazing animals. Elk migrate uphill to meadows where plants still have high nutritional value, but bison can stay in valleys like Lamar all summer due to their more efficient digestive system.

Many bison winter in Lamar, but most elk migrate to lower elevations where the feeding is better and there is less snow, often outside the park. Doug spoke of research that discovered that spring vegetation has a 12 percent protein content, but that this drops to between 2 percent and 4 percent in the winter. That is sufficient for bison, which have evolved to live in high-elevation regions like Yellowstone.

A new study by Park Service and university researchers, with Chris Geremia as the lead author, found that Yellowstone bison graze on the upper part of wild grasses, leaving the lower parts intact. This style of grazing allows the grasses to keep growing even as the bison continue to feed on them. That results in a continual supply of highly nutritious forage for the bison herds. Mark Hebblewhite of the University of Montana, one of the study's authors, told the *Billings Gazette*, "Basically, they just start lawn mowing it... and keeping it in a state of perpetual spring."

THE SEVEN LAMAR adults were in a playful mood a few days after the pup and bison incident. 06 and three of the yearlings had the fourth yearling on the ground and nipped at it. 755 ran over, jumped over 06's back, then ran around and did random leaps into the air. 754 joined in on the play and romping.

During chasing games, a pup would chase and try to tackle a sibling. That was good training for hunts where an adult wolf would have to pull down an elk from the rear. At times two pups would run at each other and at the last moment one would dodge out of the way. Experience in dodging each other would come in handy later when they were being charged by elk and bison.

The pups were already experts at following scent trails. One morning the pack traveled away from the rendezvous site. A pup got left behind, but it steadily followed the scent trail of the others with its nose to the ground, looking like a well-trained tracking dog. By that time, the pups had gained a lot of confidence, and sometimes one of them would lead when the pack traveled. All of this was practice for their future lives as adult wolves.

IN LATE SEPTEMBER, I saw 06 and 755 repeatedly chase a herd of elk that had a lot of calves. The hunt lasted thirty-two minutes. On eleven occasions, adult cows and a big bull ran over and interrupted the wolves when they were chasing a calf, preventing them from making a kill. I had seen the Lamar adults protect their pups from grizzlies, and this sighting demonstrated how elk work to protect their young from predators.

In the early hours of October 1, a grizzly broke a window on a car in Silver Gate and ate some potato chips in the vehicle. A few nights after that, I woke up at 3:30 a.m. when I heard sounds outside my cabin. I turned on my porch light and saw a grizzly running off from a pickup truck with a camper on it at a nearby cabin. The door on the camper shell had been yanked open and all sorts of items had been pulled out and strewn around on the ground by the bear. Later in the month, what was probably the same grizzly broke into three cars in our small town. I made sure not to leave food in my car overnight and never had a problem with a bear damaging my vehicle.

The Lamar wolves often traveled east from their den forest and sometimes ended up in our town. There was a legal wolf-hunting season in our section of Montana at that time. One morning I saw some of the Lamar wolves just inside the park border, slightly west of town. I then heard shooting to the east, from the direction of Silver Gate. After that, the signals from the wolves indicated they had headed west, away from town and deeper into the park. I hoped that meant 06 had some understanding that the sound of gunfire was a threat to her family.

The pack was on a fresh cow elk carcass near their den the next day. I noticed that 06 fed on the lungs and the lining of the rumen. She must have preferred those pieces, for there was plenty of meat on the carcass. I recalled the time I had seen her feed on an elk rumen and that I had thought she was doing it because the rest of the carcass was consumed. Now I wondered if those sections had nutritional components that the meat lacked. The adults were still giving the pups regurgitations, even though the five-month-old pups could pull meat off the elk on their own.

754 had been limping on one of his legs for some time by then. When it got especially painful, he held it off the ground as he walked. Through my scope I could see a swollen area just above his paw. Later I looked at a photo of that leg and saw that the paw was also swollen. Paw issues are common among wild wolves because of the extreme amount of traveling they have to do over rough terrain. Holding a paw off the ground would facilitate recovery and help prevent infection. In my experience, most foot injuries in wolves tend to heal.

NOW THAT IT was fall and the Lamar adults and pups were traveling far and wide throughout their territory, the Wolf Project wanted Colby and me to check on the den forest. We started by looking at the original Druid den first used in 1997, the site where 06's mother had been born in 2000. It was a tunnel dug through the roots of a big tree. It looked like it had been unused for years. We hiked past there to a place where giant blocks of sedimentary conglomerate rock had fallen from a cliff face below Druid Peak. Some of the massive pieces were much bigger than houses. There were many hollows and cavities under them that could serve as natural dens.

At one spot, a cave-like opening had wolf hair at the entrance. This appeared to be where 06 had her pups the previous spring. The entrance was too small for a grizzly to pass through. The passageway to the actual den chamber looked like it was about ten feet long. Nearby openings under the huge rocks were much smaller, too narrow for adult wolves to squeeze through, but accessible to young pups. In case of a major threat, they could hide there.

I had assumed 06 had used the old Druid den, but now realized she had chosen a site that offered much better protection to a litter of pups. The original den could have been easily dug out by a grizzly, but not this rocky fortress with its multiple hiding areas. I remembered how often 06 had to drive grizzlies away from her Slough Creek den site. She had done a good job at her new den site of planning ahead and anticipating threats.

IN MID-OCTOBER, I spotted a dead grizzly in Lamar Valley. I called the Park Service bear management office, and a team came out to investigate. Colby joined them as I stayed back on the road to let people know what was going on. The crew found that it was a female grizzly. She had been killed by a bite to her head by what must have been a much bigger bear. Her claws were broken, an indication she had fought back against her attacker. The strength of that other grizzly had been so great that one of his teeth had gone right through her thick skull and into her brain. The bear biologists estimated that she was fifteen years old and weighed 260 pounds.

The prime suspect was a 390-pound grizzly with fresh wounds on his face. The big male had earlier gotten into trouble outside the park when he went into a campground. Wyoming Game and Fish had captured and moved the bear to a more remote location. After release he ended up in Lamar Valley. We were concerned when we heard this was a problem bear, for he could be a danger to us and hikers, but after that incident we never heard of any encounters between the bear and people.

754 was now putting weight on his injured leg when traveling, so it must have gotten better. I noticed that he was especially alert to what was going on around him. One day, four people were hiking in a meadow near the wolves. 754 saw them first, jumped up, and ran off. That alerted the other wolves and they all followed him away from the hikers. We wanted the wolves to be wary of people, for outside the park a person walking toward a pack could be a hunter looking to shoot a wolf.

When I left home in the early mornings and drove into the park in the dark, I occasionally saw the Lamar wolves traveling down the road, going away from me. I would stay way behind them in case I needed to slow down any traffic coming up behind me. I saw a pattern in their behavior. When cars approached from either direction, the adults and pups would leave the road and slip into the thick trees, then return to the road when the car had passed through. That was the proper way for them to deal with traffic. We had Yellowstone wolves hit and killed by cars over the years, but none of o6's sons or daughters ever suffered that fate. I took this to mean that o6 was a master at teaching her pups survival skills.

NOVEMBER I WAS the first morning snow covered everything in Lamar Valley. That day we got a mortality signal from Agate male 586 toward the west end of Specimen Ridge. A Wolf Project crew hiked out and found him. There were wolf bite marks on 586 and a lot of wolf tracks around his body. Wildlife biologist Josh Irving was on that crew, and he told me the Blacktails had recently been following the scent trail of the Agates near that location. The two packs probably ran into each other and fought. 586 must have been seriously wounded during the battle.

I spoke with Doug Smith and he mentioned a statistic I had not heard: 80 percent of wolf kills in Yellowstone take place at night. I thought about how often I had come out in the early morning and seen wolves on a fresh carcass, and realized how well that fit with the 80 percent figure. This success rate also backed up my thinking that wolves have excellent night vision. Research done at the University of Wisconsin–Madison in the mid-2000s estimated that dogs can see in light up to five times dimmer than the bright light we humans need. Their pupils let in more light, they have more cones in the center of their retinas, and like many nocturnal animals they have a reflective layer known as a tapetum—the layer that lights up in your headlights at night—that acts like a mirror to improve their night vision. Since dogs are so closely related to wolves, that research helps explain wolves' ability to hunt at night.

DAN STAHLER DID a flight in early November and found the Mollie's wolves close to the Agate pack. There were twenty in the Mollie's group. Their alpha female still had not found a replacement alpha male unrelated to her. There were plenty of males in the north that could fill the role, including in the Blacktail pack. But to find prospective mates, the Mollie's females would have to trespass into the territories of wolf packs they had been feuding with for generations. A few days later, I found nine of the twelve Agates in Lamar Valley, including alpha male 641 and alpha female 715. 641 was born into the Mollie's pack, so he might still have been on friendly terms with them if the two groups ran into each other.

THERE WAS A tradition among those of us who worked in the northeast section of the park to have Thanksgiving dinner at the Yellowstone Institute in Lamar Valley. It was always hosted by Bonnie Quinn, the manager of the institute's campus, a friendly, outgoing person who made everyone feel welcome.

On the way there, I saw a man in a parking lot who looked like he was having trouble. I stopped to check on him and found that he had locked his keys in his car. Using my park radio, I called the Park Service dispatch office and they arranged for a locksmith to come out. But there would be a long travel time, several hours, and it was getting very cold. When I asked the man about himself, he said he was a veteran. I sensed he was down on his luck. Knowing Bonnie would approve, I drove him to the institute. He had Thanksgiving dinner with us, and then I took him back to his car in time to meet up with the locksmith.

AGATE ALPHA FEMALE 715 died in late November, at the age of four and a half. By that time, her pack had dwindled to eight members. 715 had wolf bite marks on her body and I suspected the Mollie's were responsible for her death. The competition among wolf packs for good hunting areas in Yellowstone was getting intense, and it seems 641's relatedness to the Mollie's had not been enough to avoid a deadly confrontation between the two packs. I did not know it then, but the situation would soon get far more intense and present a dire challenge to 06 and her family. She would need every bit of her intelligence and cunning to survive what was coming.

When the biologists examined 715, they found she had a dislocated hind leg. The other hind leg was mostly out of its socket and just barely connected by tendons. It looked like she had hip dysplasia. Those injuries would have enabled other wolves to easily catch her. She survived the attack but later died of blood loss. When I learned of her disability, I was amazed that she had been serving as alpha female. She must have achieved that position and retained it based on her relationships with other females, rather than on her ability to dominate them. Like 42, she was an example of how cooperation rather than aggression can be a successful strategy if you are a female wolf.

Doug flew in early December and saw the Mollie's wolves near another group of wolves, known as the Mary Mountain pack, in the central section of the park. The two groups soon fought and the Mollie's killed the alpha male of the other pack. Then one of the young Mollie's males took over the vacant alpha position. The males in Mollie's had the same problem as the females, in that all the females in the pack were too closely related to them to mate with. This attack on a neighboring pack allowed one of the males to pair off with a suitable mate.

I tried to figure out what made the Mollie's so aggressive to other packs. Generations had come and gone since their founding alpha male and pups had been killed by the Druids over fifteen years earlier. Perhaps over that time the Mollie's had developed a culture of classifying all other wolf packs as enemies that needed to be attacked before they could mount a surprise assault against the family, in the same way that teenagers who grow up in violent cities often act aggressively when they see strangers in their neighborhoods.

The leader of the pack, alpha female 686, had a personality much like 40, the super-aggressive Druid alpha female who gained that position by driving her own mother and a sister from the pack, then acted abusively to her sister 42 for years. Both 40 and 686 were quick to use violence to achieve their ends. With 40, that was mostly against the other females in the family, whereas with 686 it was against neighboring packs. I checked on the date the Mollie's killed the Mary Mountain alpha male and found that it was December 1. That was just five days after 686 took over the alpha female position. This would prove to be a sign of what was to come.

WE SAW NINETEEN Mollie's wolves on Specimen Ridge on December 5, just five miles from the eleven-member Lamar pack. There were now only two collared wolves in the Mollie's: alpha female 686 and younger female 779. In mid-December, we spotted the same nineteen Mollie's wolves three miles from 06 and her family. Two days after that, the Lamar wolves howled from the north side of Lamar Valley and the Mollie's howled back from the south side.

The next day the Lamar wolves were in Silver Gate, eighteen miles farther east. The howling sessions must have alerted them to the fact that they were outnumbered, so it was a wise move to slip away. I thought back to my days of watching 06's parents, alpha male 113 and alpha female 472. Both of them had a habit of avoiding unnecessary battles against larger packs.

While the Lamars were away, we got a mortality signal from Agate alpha male 641. A Wolf Project crew examined his body and found that he had been killed by wolves. The Mollie's were the prime suspects. He had been born into the Mollie's family but apparently that had not mattered. They had killed him despite the fact. Soon after that, the Mollie's wolves were at Slough Creek, where 06 had denned the previous year. That day, 06 and her family were just two miles east of there. The Mollie's were getting closer to the Lamar wolves and it seemed the two families were destined to get into a fight.

To add to the danger, the Mollie's knew where the Druid den forest was located and had often visited it in the non-denning season. During those times, the Druids had retreated to sections of their territory far away from the Mollie's and later had their pups elsewhere. If 06 were to use the Druid den forest again, the Mollie's might invade and attack her and her pups.

All of that was very troubling to those of us who had known 06 for her entire life. But we knew she was an extraordinary wolf, capable of quick thinking and skilled at surviving almost anything that came against her. Mollie's alpha female 686 had likely never encountered a wolf like 06.

PART III

2012

8

—◆—

Death on
the Lamar River

O N JANUARY 1, the Mollie's wolves were feeding on a
fresh elk kill at Slough Creek, and they were still there
the following day. The Mollie's began to howl and the
Lamars, who were eight miles to the east, howled back. The
nineteen Mollie's looked east, started to move toward the
Lamars, and howled. But then they got distracted and went
back to their kill. I later heard that 06 had been looking west
in the direction of the other wolves. She was definitely aware
of the intruders in her territory and must have been stressed
by that threat to her family.

Five days later, the two packs were on the south side of
Lamar Valley. 06's group of eleven was up high on Specimen
Ridge. The Mollie's wolves were downhill to the west. The
Lamars traveled east on their ridge. The Mollie's soon started
following their scent trail. 06's family came out on the valley

floor at the Chalcedony Creek rendezvous site. The Mollie's got confused by other recent wolf scent trails and ended up following the wrong trail south, away from the Lamar wolves. 06 had given them the slip.

MATING SEASON WAS fast approaching, and 06 and 755 started to go through preliminary courting behavior. She romped around him, wagging her tail and bumping her hip against him. Then she went over to 754, who was bedded down. It looked as though she wanted to play, but that was a deception. Instead of playing, she stole a piece of meat he had carried over to his resting spot. He jumped up and then both brothers chased her. 06 easily outran them. I got the sense that she enjoyed teasing and outsmarting them.

On January 13, I saw the Lamars sniffing around the spot at Slough Creek where the Mollie's had killed an elk on the first day of the month. As 06 did this, she would get a sense of how many wolves were in the other pack and understand that her family was greatly outnumbered.

That day, the Mollie's were to the south, up high on Specimen Ridge. Three Blacktail wolves approached from the west: the alpha pair and adult male Medium Gray. They found the scent trail of the Mollie's and followed it, howling frequently. The Mollie's would have heard their howls.

The Mollie's chased a bull elk into a gully and went out of sight. When they reappeared, I saw that nine of them were charging at the three Blacktail wolves. The Blacktails immediately turned around, ran toward the road, and crossed the pavement. Medium Gray continued north as the alpha pair swung west. The Mollie's, with alpha female 686 in the lead, pursued him. We lost sight of them heading toward the iced-over Lamar River.

Later we learned that the Mollie's had caught up with Medium Gray on the ice and killed him there. When the lead wolves came back into sight, I saw a Mollie's wolf with blood on its face. The other wolves in the attacking party reappeared, crossed the road to the south, and went uphill to join the rest of their pack.

I had known five-year-old Medium Gray since he was a pup in the Druid pack and was saddened by his death. He had approached a larger rival pack and paid for that mistake with his life. If he had been more cautious, like 06, he would have been all right.

I went to John and Mary Theberge, who were in the park studying and recording wolf howls, and we compared notes on the sequence of events. After that, they told me about some of their findings on wolf howling. John reviewed how the harmonic overtones of a howl identify an individual wolf to those that know him or her. But this section of the howl does not carry well over distances. The howling animal could be identified by others if they were nearby, but not if they were far away.

That is what must have happened when the old Agate male 586 heard the Lamar wolves howling from a distance, mistook them for his own pack, and went toward them. The Lamar wolves caught him and beat him up, then let him go. They could have killed 586 but spared his life, even though he was originally from the Mollie's pack, their traditional enemy. That incident demonstrated that 06's pack was not excessively aggressive to outsider wolves, like the Mollie's were.

Four of the collared Agate wolves had been killed by other wolves recently: alpha males 586 and 641, alpha female 715, and young male 775. The Mollie's wolves were the main suspects in three of those deaths. Plus, they had killed Blacktail male Medium Gray. That made the count four, five if the alpha male of the Mary Mountain pack was added in.

Despite all the stress and tension around her, 06 was still playful. I saw her chasing a pup back and forth until it lay down from exhaustion. After that, she ran around for no discernible reason. I had seen this behavior before, and wolves, as well as dogs, seem to do this simply because it feels good. The other wolves followed when 06 walked off. She soon dropped down into the ambush position, then jumped up and charged when a pup approached her.

We were in snow season, and early one morning drifts were eighteen inches deep in my driveway. I managed to get out to the road and the snow was much less deep once I got to lower elevations ten miles out in the park.

SEVEN DAYS AFTER the death of Medium Gray, the Mollie's were back on Specimen Ridge. The Lamar wolves were at the Druid den forest the following morning, a site they likely regarded as their stronghold. The den forest was on the opposite side of the valley and about six miles from the Mollie's.

One day we were watching the Blacktail wolves and noticed that many of them were licking each other. A young adult female licked the face of a young male, then went to alpha female 693 and licked her face. A black female licked the face of male Big Blaze. After that, Big Blaze licked the face of alpha male 778, his brother. Then 693 went to a lower-ranking female and licked the back of her neck. Five days later, I saw Big Blaze again licking 778's face. Some of that seemed to be grooming and I wondered if the behavior worked to bond the pack members together. Primate researcher Frans de Waal found that mutual grooming facilitates bonding within a group, and I felt the same thing would be true for wolves.

By that time, the Agates seemed to be down to just three members: alpha female 471, a black female yearling, and a gray female yearling. 471 looked old. Her gray coat had turned white. We had lost track of other Agates and suspected that some of them had been killed by the Mollie's. For now, it was an all-female pack. That presented an opportunity for one or more local males to join an existing pack with a high-quality territory.

ON JANUARY 24, the Mollie's were gone after having been in the northern section of Yellowstone for thirty-seven days. They likely were going back south to Pelican Valley, where they supported themselves by hunting bison. We hoped they would stay there for the rest of the winter, for that would keep 06's family safe.

I saw a spot of blood on 06's rear end on the twenty-ninth, a sign she was getting ready to breed. She did a scent mark and 755 spent a long time sniffing at the site. From that, he could get a sense of her breeding condition.

The Lamar wolves killed a bull elk at Slough Creek on the last day of the month. They had to go into the cold water to feed, and when they came out they vigorously shook their coats, then rolled in powder snow to dry off.

That day a new gray adult was with the three Agate female wolves. He looked like the Blacktail male known as Puff. He had had mange when young and lost most of the fur on his tail except for a puff of hair at the tip. Puff still had a mangy tail but had grown up and was now a big adult.

In early February, we saw two other Blacktail males with the surviving Agate females: adult male Big Blaze and yearling male 777. Big Blaze had earlier been the Agate alpha male but was displaced when two big Mollie's males, 586 and 641, barged in and joined the pack.

I thought about the deaths of 586 and 641, his successor as the pack's alpha male, and the likelihood the Mollie's had killed them. Since both grew up in Mollie's, it was a sign of how violent the pack was. If the Mollie's were responsible for their deaths, they had killed two members of their own family who had dispersed and joined a neighboring pack. It would be like a Mafia family killing two of their sons who had become members of a rival gang. With all the killing that the Mollie's were doing, I worried about them attacking 06's family.

9

The Rescue
of the 06 Female

06 HAD BECOME THE most famous wolf in Yellowstone by that time. Doug Smith had not collared her, partly owing to requests from the public who admired her beautiful light-gray coat. When he had a collaring operation on February 4, he intended to dart other members of her pack, such as one of the gray yearlings.

That morning, we found the Lamar wolves on the south side of Lamar Valley. The helicopter came in and Doug darted a gray pup and a gray adult who just stood there, looking up at him. In the past, 06 always ran into thick trees when she heard the helicopter in the distance. We all assumed the wolf who stayed in place was a confused yearling. It turned out it was 06. She was assigned the number 832, and the pup would be 820. Since she was so well known as the 06 Female, we mostly continued to refer to her by that name. She weighed ninety-four pounds and the pup was sixty pounds.

No other packs were nearby, but to be safe both wolves were loaded into the helicopter and flown to the north side of the valley, just in case the Mollie's came back. It would take the wolves a few hours to fully recover from the tranquilizer drugs. Three people from the Wolf Project staff remained on the scene, well away from the two wolves, to monitor their condition: Colby Anton, Rebecca Raymond, and Julie Tasch. I stayed down by the road to keep track of the other Lamar wolves and explain our collaring operation to visitors.

The capture crew found the Mollie's on the south side of Specimen Ridge and darted three of them. One was a female pup with thermal burns on her hind legs. She must have naively waded into a hot spring. Before darting those wolves, Doug saw the Mollie's chasing the Agate pack.

Later in the day, I saw eight Mollie's on top of Specimen Ridge and another eight down by the road below. Soon the entire pack was together again, a total of eighteen wolves. Laurie Lyman studied the adult females in the group and saw that six of them were coming into breeding condition, a stage known as proestrus. That would create a chaotic situation for the pack as the females sought out unrelated males from the nearby northern wolf packs. But many of those males came from families who had members killed by the Mollie's wolves. This meant it was going to be a complicated mating season.

The crew that stayed with 06 and 820 saw a herd of seventy-five bison heading toward the drugged wolves. Rebecca described that moment: "They weren't just meandering; they were traveling toward us with tails up with purpose!" The bison must have gotten the scent of the wolves. If they found the incapacitated 06 and her daughter, the herd would likely trample or gore them to death. The crew had to figure out how to mount a rescue. Two large boulders were seventy-five yards from 06 and the pup. That seemed to be the only place that offered some protection from the approaching bison.

Colby was the tallest member of the crew. Julie and Rebecca loaded 06 onto his shoulders. He ran through knee-deep snow to the boulders and placed 06 there. Then he ran back, got the other wolf, and raced over to the same spot. Now both wolves were next to the boulders, along with the three biologists, who stood between the bison herd and 06 and her daughter.

Colby told me that by this time the bison seemed simply curious about what to them was a very strange situation. They could see the three people clearly, but the wolves were less obvious, since they were behind the humans. Even stranger to them, as Colby said, was that since the crew had handled the mother wolf and her daughter, "the three humans smelled like wolves!" The bison herd moved closer to investigate.

Soon seventy-five of the huge animals were thirty-five yards from the crew and the wolves and still advancing. In desperation the three researchers grabbed their trekking poles and bear spray and charged forward, yelling, and waving the poles in the air, hoping that would make them look bigger and more intimidating. The crew also threw snow at the herd.

For some unexplained reason, the bison lost interest and slowly passed by the people and wolves without doing any harm. By that time, 06 and her pup were starting to recover from the drugs. The crew grabbed their equipment and headed down to the road.

Colby later told me, "Discussing the experience on the hike down, we all realized the situation was very dangerous, but our own flight response was clouded by adrenaline. Not once did we place our own safety over the safety of these two wolves. I feel lucky that I endured this experience with two other like-minded people. Many others would have simply left, putting their own safety first."

Rebecca sent me an email about the rescue and said, "It was quite a terrifying experience. Adrenaline was pumping as I thought to myself, 'We can't let the 06 Female get trampled by bison!'"

Rebecca, Julie, and Colby stood between the wolves and the big herd in that dangerous situation. All three were great heroes. Over the years, Doug Smith has recruited the best and the brightest young wildlife biologists to work in Yellowstone, and those three people were examples of that.

The following morning, both newly collared wolves were back with the rest of the Lamars. To 06, the events of the previous day must have seemed like a bad dream.

10

The Mollie's Pack Attacks 06's Family

ON FEBRUARY 9, the eleven Lamar wolves were south of Slough Creek. Ten Mollie's wolves were bedded down uphill from them on Specimen Ridge. Alpha female 686 was not with that subgroup of the pack. Later the Lamars howled. The Mollie's wolves immediately jumped up, then charged downhill with their tails raised. Soon both packs were running at each other. 06 knew from earlier group howls that the Mollie's pack was much larger than her family, but could not yet tell that only ten of the nineteen were present. The Lamar wolves scattered in different directions, the proper strategy when a pack thinks it is greatly outnumbered.

There was a confusing melee of twenty-one wolves running back and forth. A Lamar black pup got too close to eight

Mollie's. They caught and pinned it. The group attacked the pup for about a minute, then got distracted by Lamar wolves running past them. The Mollie's left the pup to chase them. They failed to catch any of the Lamars, so the Mollie's raced back to the pup, probably intending to finish it off.

Josh Irving saw that pup get up after the attack and run off. The Mollie's picked up its scent trail, ran along it, but failed to find the pup. We spotted it running toward the road. The pup reached the pavement, ran east along it, then veered off to the northeast. I last saw it heading toward Lamar Valley. The Lamar River and the park road were now between it and the Mollie's, so the pup seemed safe from further attack. When it was on the road, people saw blood on its coat.

We lost sight of the Mollie's south of the road and heard that the main group of Lamar wolves was north of the road. We all spent that night worried that the Mollie's might cross the road in the dark and attack 06 and her family. I was getting ready to go back out early the next morning when local wolf watcher Doug McLaughlin called to say that he had just heard wolves howling close by.

I went outside in the cold and did a signal check. I got signals from all four Lamar wolves to the south of my cabin. I jumped in my four-wheel drive and drove out on a snow-covered back road. I soon picked up wolf tracks going east along the road. From the place where the confrontation had happened the previous day to this road was a trek of twenty-three miles. It was a smart move on 06's part to put that distance between her family and the Mollie's.

I headed out to the park and heard from Colby Anton that he had found a black pup in Lamar Valley. I spotted the pup a few minutes later, north of the road. People who saw it earlier told me the pup seemed stiff when it moved around, and that it had puncture wounds on its body. This had to be the pup that had been attacked yesterday. After watching it for some time, I got the sense that its wounds were not life-threatening.

FOUR DAYS AFTER the attack on the pup, I saw the Lamar pack near their den. I counted ten in the group, including all three black pups. The wolf family moved off and I saw that the black pup who had been attacked was last in line. It was moving like it was in pain but stubbornly kept up with the others. 06 would have sniffed that pup when it rejoined the pack and gotten the scent of the Mollie's wolves where they had bitten her pup. I was sure that she would never forget that scent. I recalled how wolf 42, 06's great-aunt, had to deal with the violent personality of her sister, wolf 40. Now 06 would have to deal with the equally violent 686.

Soon after that, I saw a pup stop at a rosebush and eat the fruit on it, known as rose hips. A second pup came over and both fed on the bush. That was the first time I had seen wolves eat plant material. I later talked to a wolf biologist who told me he had found many wolf scats in western Montana that were full of huckleberries. Then I learned that a researcher in Voyageurs National Park in Minnesota had seen a mother wolf regurgitating blueberries to her five pups. Blueberries are in the same genus as the western huckleberry, so the Montana and Minnesota wolves were consuming similar berries. The researchers at Voyageurs estimated that blueberries make up as much as 83 percent of the July wolf diet in that park.

THE MOLLIE'S WOLVES continued to pose a threat to the wolf packs in the northern section of the park. Nineteen Mollie's were seen chasing five Blacktail wolves on Specimen Ridge. Soon after that, they went back to their home base to the south, in Pelican Valley. That calmed things down in the north, at least for a while.

But for the remainder of the winter, the Mollie's would spend far more time in the Northern Range of Yellowstone than they had in past years. They would come north, stay for several weeks, return to Pelican Valley, then reappear. Because there were so many wolves in the pack, it was hard to tell if alpha female 686 had recruited a mate, but it appeared that she had not. Thinking about her situation, I wondered how she could find a male who would risk pairing off with a female with such a violent streak. What would she do to one who displeased her?

IN MID-FEBRUARY, A bison was killed in a collision on the road in Lamar Valley. Wolf Project crew member Rebecca Raymond and several patrol rangers dragged the big carcass off to the south so the wolves could feed on it. Later 06's family found that bison and fed on it. She and 755 mated multiple times over the next three days, and we calculated that her due date would be around April 20.

Back at the Agates, Big Blaze, the Druid wolf who helped start the Blacktail pack, mated with alpha female 471. Later he went to a black female and licked her ears. Her fur was sticking out at one spot on her coat and we wondered if she had been bitten there by a Mollie's wolf. She walked off with a limp, another possible injury. Then the male licked 471's back. We saw that 471 had bite wounds on her face, also probably inflicted by Mollie's. A few days later, she mated with Puff, another former Blacktail wolf. Big Blaze saw what was happening and ran over. He pinned the younger male but could not stop the mating. Right after that, Big Blaze got into a mating tie with the black Agate female.

It was time to do more radio collaring, but for this operation Doug Smith brought in a helicopter that used net guns to capture elk and wolves. That crew got fifteen elk, then netted Big Blaze and 471. He got his first collar, and she got a new one. From now on, he would be known as wolf 838.

By February 22, 06 must have been out of season, for 755 showed no response when she flirted with 754. Kirsty Peake, an animal behaviorist from Devon in the UK, was in the park, and I asked her about mating behavior in female dogs. She explained that their estrous cycle is about twenty-one days long, and that they can get pregnant from around day seven to day fifteen. Generally, they do not want to mate before or after that period. Females have blood on their rear end during the early part of that cycle, then are ready to breed when the blood is gone. From what I had observed over the years in Yellowstone, Kirsty's information seemed to apply well to wolves.

I felt sorry for 754. Wolf 755 kept him away from 06 during the mating season. I saw 754 try to get Middle Gray interested in him, but she stood him off. She was probably his niece, so she was too closely related to him for breeding. I was always impressed with how big males like 754 accepted rejection from females during the mating season. I have never seen a male wolf try to force a female.

IN EARLY MARCH, the Lamar wolves were in Round Prairie. One of last year's pups was south of the road and the rest of the family was out of sight in trees to the north. The pup howled as it attempted to locate them. The others howled back, and the pup crossed the road and headed directly toward the sound of their howling. It correctly figured out exactly where they were from the howls.

I got signals from 06 at the Druid den forest on March 4, a sign she was intending to have pups there again. The pack later came into sight and I saw that 06 was especially playful with 755. That play would strengthen their emotional bond and likely motivate him to work hard to support her and the new pups.

A few days later, the Mollie's were back in Lamar, up high on Specimen Ridge. They were down in the valley the next morning. I saw 06 and her pack sitting up north of the road, looking down at the rival wolves. They did not howl at the Mollie's. That would have given away their location to the much larger pack. Later in the day, when the Mollie's were far away and probably out of hearing range, the Lamars howled back and forth with the Agates, a pack that was smaller than theirs.

06's new collar had GPS capacity and periodically transmitted her locations to a satellite. We could download that data, and Doug Smith would later send out a crew to places her pack had been. That enabled the Wolf Project to look for kills we otherwise would not have known about. For example, a mule deer carcass was found up in the den forest and a bull elk in the trees near Round Prairie. All of that allowed us to document a more complete account of what the wolves were killing and the age and physical condition of those animals.

The snow was very deep in mid-March. I saw 06 struggling to plow through a drift. It was too exhausting for her and she turned back. 755 found a route with less snow, and 06 and the others followed him in single file. Breaking trail is one way big males in a wolf family serve the pack. During those weeks, 06 often left the other wolves and went to her den, probably to clean it out and make sure no rival wolves had found it.

THE MOLLIE'S WERE still in Lamar Valley. One day I spotted them chasing a herd of bison running single file through deep snow up a ridge. When they reached the last animal in line, the wolves attacked her, but another bison ran back and drove the pack off. After that, the Mollie's went after a bison calf. The mother charged at the wolves, hit one with a lowered head, and tossed it up into the air. The wolf landed in soft snow and got right up, seemingly unhurt. Another cow bison ran in and helped the mother drive off wolves from the calf.

Local artist and sculptor George Bumann had hiked up on a ridge to the north. He called down to say he had a good view of the wolves and another bison herd from his position, so I trudged up through the deep snow and joined him. We could see all nineteen Mollie's wolves across the valley. They went to a cow bison the pack had injured prior to my arrival, but had had to leave when fifty bison approached to protect her. By the time I arrived, only four bison were still with the injured cow. Those bison charged at the wolves, then seemed to lose interest in guarding her. The Mollie's walked up to the cow and bit her a few times. She must have already lost a lot of blood, for she was soon dead.

Seeing the Mollie's wolves successfully hunt bison that day helped me understand how their ancestors, going all the way back to their original alpha female, wolf 5, figured out how to kill bison in Pelican Valley after they were driven out of Lamar Valley by the Druid wolves. I thought back to that 1996 incident. The likely instigator of the fatal raid was Druid alpha female 40, who was well known for her aggressive personality. The violent multigenerational feud between the Mollie's wolves and 06's genetic line was now in its sixteenth year. I thought of the section of the Old Testament that speaks of visiting the sins of the fathers upon their children, to the third and fourth generation. In this case, if the Mollie's attacked the Lamar den they would be visiting the violent behavior of 40 upon her innocent granddaughter.

A Canada goose died along the Lamar River near 06's den. She came down and ate it. The pack apparently had not made a kill recently, for that day two of her pups scavenged on a bull elk that had died nine months earlier. 06 needed to eat well to support her pregnancy, but last year's pups could get by on less nutritious fare.

At the end of March, Mollie's alpha female 686 and the other pack members were seen back at the family's traditional denning site in Pelican Valley after spending seventy-six days up in the northern part of the park. If 686 was pregnant, the pack would have to stay in Pelican Valley for many months, leaving 06's family safe during the denning season.

11

The Den Raid

AT THE START of April 2012, the Lamars had ten members. Young males Dark Gray and Light Gray had gone to seek out mates, leaving the three founding adults, two young female adults (776 and Middle Gray), and the five pups from the previous year, who were now yearlings. One of those black yearlings soon disappeared, bringing the pack count down to nine.

On the morning of April 1, I spotted 06 running east on the road, looking to the south. I scanned that way and saw fifteen Mollie's racing toward her. When cars got in their way, the Mollie's turned back. I wondered whether 06's actions were deliberate. She was used to dealing with the road and traffic and may have guessed the Mollie's would be too wary of those things to continue their chase. There were no roads in their remote territory. As they retreated, I saw 754 calmly leading all the Lamars, including 06, toward the den forest.

The following morning, Rebecca Raymond found the Mollie's at Hellroaring Creek. I called Dan Stahler and he said that during recent tracking flights no carcasses had been seen anywhere near the Mollie's den in Pelican Valley. The pack must have come back to the north to find better hunting areas. Since it was April and pups were normally born later in the month, this trip north, so far from their normal den site, probably meant that 686 had failed to find a mate and was not pregnant. If that was the case, the pack might stay up in Lamar Valley for a long time. That would be bad news for 06 and her family.

I checked on what we knew of pup production in the Mollie's pack from the time 686 was a yearling in 2009 through 2011, the last year 486 was the alpha female. There were twenty pups born during those three years, and nineteen of them survived to the end of the year they were born, an exceptionally good record. But this year, 686's first as the pack's alpha female, no pups had been seen.

I joined Rebecca at Hellroaring and saw the Mollie's feeding on a bison carcass. A man in a nearby parking lot had seen the wolves kill the bison the previous evening. We got a count of seventeen Mollie's near the carcass. They stayed there over the next few days, far from the Lamar wolves.

On April 11, the conflict between the Mollie's wolves and the northern packs got more intense. Josh Irving from the Wolf Project found Agate alpha female 471's body near Tower Junction. I joined him and others for her necropsy at park headquarters. She had wolf bites all over her. There were seven fetuses in her uterus, each about two to three ounces. They were just barely recognizable as wolves. 471 weighed ninety-three pounds, about the same as her sister 06, and had a lot of fat on her. That meant she had been in good shape for her pregnancy. Her hind legs were thickly muscled. We figured she had run into the Mollie's wolves.

I saw 06 south of the road soon after that and she looked very pregnant, so much so I was surprised she had walked that far from her den. 755 was near her, but 754 was some distance away. 06 raised her head and howled. 754 looked in her direction, then ran toward the sound of the howling. When a yearling howled from another direction, 754 stopped and looked that way, then continued to run directly toward 06, even though he could not see her. I took that to mean he could tell 06's howl from the yearling's. This confirmed what John and Mary Theberge had told me about the harmonic overtones in wolf howls enabling other wolves to identify individuals if they are not too far away.

It had been a long winter, so it was joyful to see flowers that day in two locations in Lamar Valley. Buttercups were usually the first ones to bloom.

APRIL 20 WAS 06's estimated due date. We saw most of her family in the den area that day and I guessed they understood she was about to have her pups. When I went out early two days later, I saw that the pack had a fresh bull elk carcass just across the road from the den forest. 776 was bedded down, and when she got up I saw that her hindquarters were stiff. She stood in a hunched stance. The bull must have kicked and injured her during the hunt.

I checked a few hours later and saw 06 at the carcass. She no longer looked pregnant and had signs she was nursing pups. After feeding she went right back to the den. Newborn wolf pups are unable to keep themselves warm on their own, so the mother cannot leave them for long. When 06 rejoined them, the pups would have cuddled up to her to be warmed by her body heat.

Laurie Lyman and her husband, Dan, saw sixteen Mollie's wolves at Slough Creek early on April 25. None of the females looked pregnant and as far as we knew none were denning. I soon saw them north of the road in Lamar Valley, traveling east. They were moving quickly, following a scent trail likely made by Lamar wolves heading to the family's den. I began to worry about 06. The Mollie's had killed 471, 06's sister, earlier in the month and now it looked like they were coming for 06.

Wolves normally bed down in the midday hours, and the Mollie's did that at 1:06 p.m. They got up five hours later and continued east. At 7:06 p.m. I saw them on a ledge just uphill from the den forest. There was a trail from the ledge that led down to the den forest, and the sixteen wolves, including 686, went along it. As I lost them trotting downhill into the trees, I thought about the two times 40 had marched toward 42's den and killed her pups.

I realized that the Mollie's were close to where their ancestors had denned in the spring of 1996. That time, it was the Druid wolves who were on a den raid; sixteen years later, the Mollie's were the aggressors. We knew or suspected that the Mollie's had killed at least six local wolves during the previous five months and another one elsewhere in the park. I had to assume that they now intended to attack the Lamar pack.

I WAS IN the Hitching Post parking lot across the road, south of the den. As soon as I lost sight of the Mollie's, I took out my telemetry equipment and got signals from 754, 755, and yearling female 820 to the north. There was no signal from 06, and I took that to mean she was in her den.

Laurie Lyman and filmmaker Bob Landis were with me and we all knew the sixteen Mollie's were almost certainly going to hunt down 06's family. We walked out on some hills to the south to get a better view of the den forest. Bob spotted the Mollie's going through a break in the trees. They were headed directly toward the den.

I visualized the Mollie's wolves running through the den forest, encountering 754 and 755, and attacking them. It would not be a fair fight, since it would be sixteen against two. 06, down in the den with her newborn pups, would hear the battle and protectively put herself between her pups and the den entrance, ready to fight to the death to protect her litter.

Everything was quiet for a few minutes, then seventeen wolves exploded out of the trees. A collared gray was out in front. It was 06, and the Mollie's pack was chasing her. 06 was coming from the section of conglomerate rocks where she had denned the previous year, and I thought that in desperation she was trying to lure Mollie's away from her newborn pups.

As she had recently given birth, 06 could not run fast. The other wolves were gaining on her. That was bad enough, but she had another major problem. She was running toward the top of a cliff. She would have to stop there and turn to face the sixteen wolves. She was a good fighter but no one wolf could win against that many opponents. Those of us watching 06 were too tensed up to say anything. I had known her since she was a pup and was about to witness her violent death.

But I had greatly underestimated her. 06 did not stop on top of that cliff. Instead, she raced down through a steep gully strewn with sharp-edged loose rocks. I cringed as 06 ran through those rocks, for I imagined that they were cutting into the pads on her paws. In a few moments, she reached the road and crossed it to the south, a few hundred yards west of us.

I looked back uphill. The Mollie's did not want to risk running down that dangerous route. From their position at the top of the cliff, they saw a Lamar gray female to the east and chased her instead. It was Middle Gray, one of 06's two-year-old daughters from her first litter. I had previously seen that Middle Gray was the fastest wolf in her family. She ran off into the forest farther to the east, and all the Mollie's went after her. I lost the pursuing wolves in the trees.

We later saw Middle Gray well east of the den forest. She repeatedly gave a high-pitched yipping alarm call to the other Lamar wolves. People down the road told me the Mollie's had given up on trying to catch the young female. They had turned around and gone back into the forest, heading toward the spot where I was getting signals from the two Lamar males.

I heard that 06 had recrossed the road and gone back uphill. I saw her and yearling 820 on the slope west of the den forest. They were looking intently down toward the den. The Mollie's were now between them and the family's pups. It was 7:38 p.m.

06 had done her part in trying to lure the rival wolves away from the den, as had her daughter. It was now up to

754 and 755, for I realized their signals were coming from the place where I had seen 06 run out of the trees. She must have been running from her den, so the males were likely stationed there. This was going to be the ultimate test for the two brothers. The fate of the pups was on them. Would they abandon the litter and run away or stand their ground?

I realized that 686 had picked the perfect time to attack the Lamar wolves, for 06, the leader of the pack, was at her weakest after just giving birth to her pups. It was a brilliant strategy.

For thirty long minutes, we had no idea what was happening at the den. At 8:08 p.m. I saw all the invading wolves leave the den forest. Later I spotted them farther away to the west with alpha female 686 in the lead. The raid was over. We did not know what had happened but feared the worst. How could 754 and 755 have survived an assault by sixteen aggressive wolves with a history of killing a lot of other wolves? Thirty minutes was plenty of time for the Mollie's to kill the two males, get into the den, and slaughter the pups.

We then saw Lamar female 776 howling from south of the road, and I got signals from yearling 820 in that direction. A black Lamar yearling was spotted, and we heard Middle Gray howling from the east. I checked on 754 and 755 and got loud signals from them, still exactly toward the den.

An hour later, I went west and saw the sixteen Mollie's cross the road to the south. They continued in that direction and seemed to be leaving the valley. I returned to the Hitching Post lot and did not get any signals from 06, which meant she had likely gone into her den. The signals from both males were still coming from that same section of the den forest. I took that to mean they had not abandoned their defensive position. They were normal signals, rather than the mortality mode. Those of us on the scene could now let go of some of our tension. It looked like the Lamar adult wolves had survived the attempted assault. But what about their pups?

A disturbing thought then came to me. The electronics in a radio collar are programmed to wait four hours once no movement is detected in a wolf. Then the collar doubles the beeps per minute, notifying researchers that the animal has likely died. I saw the Mollie's leaving the den area at 8:08 p.m., so if they had killed 754 and 755, the mortality signal would not kick in until close to 12:08 a.m. The signals I was getting now could be deceptive. The two brothers, along with all the pups, might already be dead or close to death.

As I drove home, I felt like I had been functioning as a war correspondent over the past few hours, one who did not know the outcome of a great battle but suspected that some of the soldiers he had known and admired had been killed.

I RETURNED EARLY the next morning. Fearful about what I might hear, I carefully tuned my receiver to 755's frequency. I got a normal signal from him toward the den, then got the same from 754. I now knew that they had both survived.

Then I remembered something. Back in 2010, when the Lamars had their first litter of pups at Slough Creek, there was a day when a grizzly approached the den. I saw 755 charging at the bear, then noticed that 754 had positioned himself at the den entrance, ready to defend the pups inside if the bear approached.

I pictured the two brothers doing something similar when the Mollie's approached this den, standing shoulder to shoulder, blocking the entrance from the sixteen rival wolves, willing to fight to protect the pups. Then I wondered: what level of courage does an individual need to have to do that?

While thinking about this, I came across a quote from President Franklin Roosevelt regarding the start of World War II and the fear Americans had that we could not prevail against that existential threat to our country. Roosevelt said, "Courage is not the absence of fear, but rather the assessment that something else is more important than fear." For those brothers, the "something else" was 06's litter of newborn pups.

06 had been courted by many large and experienced male wolves but chose two yearlings to form a pack with her. I always wondered what she saw in these young and untested brothers. I believe 06 sensed she could count on them to do their duty to protect their family, regardless of the danger. In all aspects of her life, 06 was a superior wolf, and I now saw that her ability to judge who would be the best wolves to start a family with had enabled her to make an exceptionally good choice.

Dan Stahler and I later talked about the analysis of the DNA samples from 754 and 755 and how they were related to the Mollie's wolves. Their connection to the Mollie's intrigued me. When that pack of sixteen charged into the Druid den forest, the two brothers were standing up to their own relatives.

I DID NOT get a signal from 06 that morning but thought that indicated she was still in the den. I then went west and saw the sixteen Mollie's hunting bison. They howled and the Lamars defiantly howled back in response.

Over the next few days, the signals from the Lamar wolves continued to come from their den forest. 06 seemed to be going in and out of her den like a normal mother wolf with a healthy litter. She was seen leaving the den forest on April 29, four days after the Mollie's invasion. Soon after that, she returned. This was another indication everything was all right at the den and with her pups.

Later we saw 755 and Middle Gray leave the den and come back. I got a good look at them and they seemed unharmed. I added up that we had seen all the Lamars except for 754 and one yearling since the den assault. 754's signal was coming from the den. I could tell that he was moving around in a normal manner, so it seemed he also was fine. I had previously seen that 754 was especially attentive to the pups and liked to be with them. I figured he stayed with the litter as the other adults went back and forth. Having the big male with them would have calmed and reassured the pups after the traumatic attempted raid on their den. By that time, the Mollie's wolves had left the Lamar pack's territory.

In past denning seasons, it had taken many weeks for the young Druid pups to leave the area immediately around their den and start to explore the surrounding forest and meadows. We would have to wait patiently until this year's litter finally came into view and we could count them. 06 had four pups in her first litter at Slough Creek in 2010 and five here in 2011. Hopefully, we would get a similar count this year.

An Alliance of Sister Wolves

There had been another den raid prior to the one the Mollie's mounted on 06's family. Back in 2006, it looked like four females in the eleven-member Slough Creek pack were pregnant and were going to have their pups at their traditional den site on a ridge west of Slough Creek. There were three big males in the family at the time.

A new pack came into the Slough territory from north of the park. Since we did not know them, we called the group the Unknown pack. There were twelve adults in the pack. The new group killed one of the big males in the Slough pack, then laid siege to their den. The two surviving Slough males, including alpha male 490, seemed to give up on defending their family. They left the den region and took up with a new female. Their abandonment of their females and pups astounded me, for I thought that all alpha male wolves would be willing to fight to the death to protect their families.

The Slough females had no choice but to stick together and protect the den and the pups inside from the rival wolves, who often were camped out just a few hundred feet from the den entrance. I would see individual Unknown wolves try to enter the den tunnel, then jump back, acting like a female inside had just bitten them. It looked like there was always a guard there, risking her life to block any intruders from getting the pups at the far end of the tunnel.

Day after day the siege went on. I felt like I was watching a medieval siege of a castle. At times one of the mother wolves would sneak out to get a drink of water or search for food. I saw two of them leave, then run back when the rival pack spotted them. They got into the tunnel a split second before the Unknown wolves reached the entrance.

Younger females in the pack also risked their lives to help the mothers and pups by going out for food and trying to bring it back to the den. The siege went on for thirteen days. None of the pups survived but all of the females did. The mother wolves just did not get enough food and water to keep the pups alive.

Witnessing that siege gave me great respect for female wolves and the tremendous courage and determination they have when it comes to protecting their pups. The males in the Slough pack were all older and bigger than 754 and 755. They abandoned their females and newborn pups to save themselves, while the two brothers stood their ground. That comparison gave me more reason to be impressed by 06's selection of 754 and 755 to help her start a pack. She knew what she was doing.

12

Teamwork

THE MOLLIE'S STAYED in the north during May, but there were no more encounters between them and the Lamar wolves. No unrelated male had come into the pack, so alpha female 686 still did not have a mate. As of May 2, the pack had been in the north for one hundred days.

The Lamar pack now had nine adult members. Middle Gray was the second-ranking female in the pack. There were three yearlings: one gray female and two blacks. The two blacks were hard to tell apart, but one was male and the other female. I gradually found some minor difference in their coats. The female had darker fur than her brother. She would later be collared and designated as wolf 926, so I will call her that from now on. 926 liked to dig ground squirrels out of their tunnels and eat them. She was the black pup that had stood out to me the previous year for her leadership qualities.

Erin Albers, who had recently married Dan Stahler, did a tracking flight on May 6 and got the Lamar signals from the rocky area on the back side of the den forest. When we next saw 06, it was obvious she was nursing. I took that to mean her pups had survived the Mollie's assault. But I needed to see the pups to be sure.

I normally lived in wolf world, but sometimes, for a few hours, I would leave Yellowstone and visit what some people consider the real world. Wildlife guide Carl Swoboda and I went to Bozeman, Montana, to see the first Avengers movie. The way the superheroes operated as a team reminded me of how wolf packs such as 06's family functioned, particularly during the den raid. When doing talks to elementary school groups, I would sometimes joke with them that Tony Stark had once come to Yellowstone to recruit 06's famous grandfather, 21, as a member of the Avengers.

THAT SPRING A bison calf got separated from its mother southeast of the den forest. It stayed by Soda Butte Creek and constantly called out for her. The gray female yearling from the Lamar pack happened to walk by and the calf ran toward the wolf, perhaps thinking any moving animal would be its mother. The young wolf first seemed confused by a prey animal running to it. Then she bit the bison calf, but because of the calf's thick hide little damage was done. The young bison called out in distress.

A cow bison was nearby. She must have been the calf's mother, for she charged toward it. The wolf let the calf go and trotted off to avoid the rapidly approaching mother, but the confused calf followed the yearling. Before it could reach the wolf, the mother bison ran in, passed by the yearling, and reunited with her calf. She sniffed it and the calf immediately started to nurse. The wolf bedded down and calmly watched the pair she had unintentionally reunited.

On May 1, Kira Cassidy, a Wolf Project biologist, drove up to Antelope Creek, six miles south of Tower Junction, and got a mortality signal from male 838, also known as Big Blaze. Although he had once been the Agate alpha male, by that time he was back with the Blacktail pack. Kira spotted his body in a meadow a mile east of the road.

I went up there and joined her. We saw twelve Mollie's wolves near that meadow, including violence-prone alpha female 686. Kira had gotten signals from another Blacktail wolf in the area: male 777. He had a GPS collar, and when the location data was downloaded, they showed that he had been at Antelope Creek at 3:30 a.m., exactly where 838's body was found. 777 must have been checking on his packmate.

A later examination found 838 had died from wolf bites. The Mollie's wolves were the likely killers. We were thinking that the Mollie's aggression against the northern wolves had ended, but the death of 838 proved we were wrong. He was the seventh wolf they had killed in our section of the park.

After that, things got complicated. We eventually found that a contingent of Mollie's females and a few young males had a den at Antelope Creek, but the mother wolf was not 686. Former Blacktail male 777 and his packmate Puff were also in that group. At times we counted six adults in what seemed to be a new pack. A big gray wolf appeared to be the alpha female. I thought about how ironic it was that 686 had never found a mate up in the northern section of the park but some of the younger females in her family had.

ON JUNE 22, Dan flew over the Antelope Creek den and saw two black pups. By that time, most of the other Mollie's, including 686, were back in Pelican Valley. Doug did a later flight and spotted three black pups. He also noticed that Puff was acting as the top male in the group.

There was bigger news from that flight. As Dan circled over the conglomerate rocks on the back side of the den forest, he spotted two black pups and two gray pups. We finally had proof that the two Lamar males had successfully protected 06's pups during the Mollie's den raid two months earlier.

I recalled the time when Colby Anton and I investigated the den forest. We found the den 06 had used in the spring of 2011 under huge rocks that had fallen from a high cliff. The back end of that cave-like den was very narrow and nearby openings were even smaller, too tight for an adult wolf to enter. When the Mollie's were in a standoff with 754 and 755, the frightened pups could have squeezed into the back reaches of one of those areas, places that amounted to safe rooms. If a few of the sixteen Mollie's wolves had made an end run around the brothers and tried to get into any such sites, the pups would have been beyond their reach. I gave credit to 06 for great intelligence and foresight in choosing that location to den. The combination of the brothers' defense of the site and the hiding places under the rocks enabled the pups to survive the onslaught of the rival wolves.

ONE MORNING I was watching 06 and four of her adult daughters. 754 approached the group and the five wolves ran to him and greeted him in a very friendly manner, like a human family hugging their favorite uncle. 754 seemed to have an easygoing personality that caused the other wolves to like to be around him. I saw him lick the face of young female Middle Gray and thought he meant that as an affectionate greeting.

In mid-July I saw another side of 754. The three adults and three younger wolves were at the edge of the Lamar River, watching a bull elk standing in fast-flowing water. They must have chased him there. An experienced elk will often run to deep water when pursued by wolves. With his long legs, the bull can stand in the river, but wolves have to swim out to confront him. That puts them at a significant disadvantage, for they lose most of their leverage if they attack their opponent while treading water.

I saw 06 slip into the river and dog-paddle toward the bull. 755 followed her lead. Both wolves had to swim against the fast current to reach the elk, who was confidently standing his ground. As they closed in, he suddenly ran through the water downstream past them. The wolves turned around and swam after him. When they got close, he turned again and ran upstream past them. The alpha pair pivoted and struggled against the current as they swam after the bull. He could travel upriver much faster with his long strides than they could manage by swimming.

At one point, the bull stood in shallower water and 06 charged at him. She leaped up at his throat, but he nimbly dodged out of her reach. 755 tried to grab a leg when the bull ran by him through the water and missed. During all these attempts, 754 and the other Lamar wolves just watched the alphas try to get the elk and did not go in the water.

755 then tried a new strategy that impressed me for its cleverness. He got out of the water, trotted upstream past

the bull, then slipped in the river and floated down toward him. The big elk turned to face the wolf, reared up, and kicked down at him. 755 dodged the blow. He then got a bite in on the elk's shoulder but was shaken loose.

754 and a gray female both briefly got in the water and tried to help out but, like the alphas, were outmatched by the bull's ability to stride up through the fast current. Both quickly gave up and came out of the river.

The alphas must have been exhausted by then but kept trying to get the elk. 06 leaped up at his throat and missed once more. 755 got a bite on the bull as he passed by but lost his grip. 06 would periodically come out of the river, roll in grass to rub the cold water out of her fur, then go back for another try.

After twenty-eight minutes of intense, exhausting attempts on the elk, the alphas came out of the water for a rest. Forty-eight minutes later, 06 got up and slipped back into the river. 755 joined her. The wolves must have sensed that the bull had some weakness that would eventually allow them to finish him off. Their epic battle with the elk was about to continue.

755 got a bite on the bull's hindquarters but soon lost his hold. 06 joined him and bit into the left side of the bull. That slowed the elk down and enabled 755 get a good hold on his shoulder. Thinking quickly, 06 let go, swam forward, and grabbed the side of the elk's neck. Working together, the two wolves, who probably had a combined weight of about 200 pounds, pulled down the bull. He looked to be about 500 pounds.

At that critical moment, 06 lost her grip and the bull was about to break free. All the intense efforts the alphas had invested in the hunt now seemed to be wasted. I had forgotten that 754 was on the scene. He must have seen what was happening, for at that critical moment he ran into the water, swam over to the alphas, and got in front of the bull. Leaping out of the river, 754 got what looked to be a fatal bite on the throat. His actions inspired three younger wolves to jump in the water and help. All six wolves had to tread water as they jointly attacked the elk. I saw the bull's head slowly drooping into the water. The younger wolves swam away at that point, leaving the alphas and 754 to continue the attack. Their holding bites were forcing the bull's nostrils under water. That did it: he died by drowning. The fight with him had lasted eighty-two intense minutes.

The wolves pulled the carcass to shore and began feeding. That hunt demonstrated the importance of teamwork in a wolf pack. The alphas took the initiative and did most of the work. But they desperately needed help at the end. That was the moment 754, who was the biggest and strongest wolf in the group, proved his worth. He made the killing bite. 06 put together a team when she recruited those two brothers years ago, and this hunt was more proof of the efficiency of that team. 06 fed for over an hour. Then, when she was full, she crossed the road and went to the den to regurgitate the meat she had eaten to her four pups.

Later in the month, long after the carcass had been fully consumed, I hiked out to examine the place where the bull had been killed. Both sides of that section of the river were covered with rocks the size of soccer balls that shifted under my feet as I stepped on them. It was like trying to walk on a field of giant marbles. The surface would have made it hard for the wolves to keep their footing as they raced up and down the river after the bull. There was a three-hundred-foot section that had the fastest current and deepest water. That was where the bull had run back and forth, up and down the river. The elk had chosen a strong defensive position, but the persistence of 06 and her family overcame the advantages on the bull's side, and he lost his epic battle with them.

I saw 06's four new pups up at the den for the first time soon after that. They would have been about three months old by then. In my next sighting, I saw the pups walking around in single file. The female black yearling 926 was supervising the litter.

PART OF MY job with Yellowstone's Wolf Project involved helping park visitors see wolves. Usually that was a simple process. I would look for wolves from one of the parking lots along the park road. If I found them, I would let people see them through my high-powered spotting scope. Sometimes I needed to walk uphill to find wolves that were not visible from lower levels. In those cases, people along the road would come up and join me.

One day a man in a wheelchair was in one of our lots when I arrived. He told me that his dream was to see wild wolves. We could not find any wolves from the parking lot, so I hiked up the slope to the north and soon found a pack. The problem was that people had to come up the hill to see the wolves. I went back to the road and asked several of the longtime wolf watchers to help me do something. We pushed the wheelchair uphill and as the man locked it in place, I set up my scope in front of him. The wolves were still there and he got to see them.

AT THE END of July, I was in Round Prairie, five miles east of the den forest. I heard a sound and saw a cow elk running west along the banks of Soda Butte Creek. 06 was chasing her. The cow veered into the water and continued to run. The wolf followed her into the creek and had to swim when the water got deeper. 06 caught up with the elk, leaped up, and bit into the side of her neck. The cow bucked up and down, broke free, jumped over 06, and ran off.

I lost them behind some trees, then the cow reappeared, running through the creek, back to the east. 06 was right with her, sometimes swimming, sometimes running. She reached the cow and bit into her back. But 06 lost her grip and, now in deep water, had to swim after the cow. At that moment, the elk turned around and charged at the wolf. Undaunted, 06 leaped up out of the water, aiming at her opponent's throat. Dodging the bite, the elk continued to plow her way through the water. Despite the elk's advantage in the creek due to her long legs, 06 caught up and grabbed her neck. The elk bucked up and down as she continued running. The wolf held on for a bit longer before being shaken off. The two ran west through the water and I lost sight of them. We got glimpses of 06 still on the attack and soon we saw that the elk was dead. That incident convinced me more than ever that if there was ever a female wolf who had the spirit of an Amazon warrior, it was 06.

Jeremy SunderRaj, a high school student who had a summer job in Silver Gate, was with me and he took a lot of photos and a video of the chase and attack. Jeremy had started coming to the park to watch wolves when he was ten years old. He later got a degree in wildlife biology at the University of Montana and joined the Wolf Project staff.

We repeatedly played the video to study the details of what 06 did. It showed the moment when she leaped out of the water toward the cow's throat. The elk turned her head, causing the wolf to miss her. Later in the footage, as the cow ran away through the creek, 06 veered out of the deep water to a shallower section and that enabled her to gain on the cow. At one moment, as she ran beside the elk, we were struck how small 06 looked compared with the cow. Despite that, she killed the much larger animal.

Other Lamar wolves came to feed on the dead elk. We had to put Closed signs at Round Prairie and No Stopping signs on the road so people would not gather and keep the wolves from feeding. When I later saw 06, I noticed that she held her hind right leg off the ground when walking. I could see that it was swollen. The elk had probably kicked her there. 755 carried off a leg, left it on the ground some distance away, and returned to the carcass. One of the younger wolves who was wary of people had not yet fed. She went right to that leg and ate. I wondered if 755 had deliberately carried it there for the shy pack member.

I had not seen 754 near the carcass or picked up his signal, so I drove to the den and got it there. He was taking care of the four pups while the alphas and the other six adults were out on the hunt. 754 came to the carcass in late evening and got his share of meat then.

I kept records of successful elk hunts by wolves I witnessed. That cow was the 133rd elk kill I had seen. This was my 5,447th day out in the park since the wolves had been brought back. I did some math and figured out that I had seen wolves kill an elk an average of once every forty-one days. Since most successful hunts take place at dark, I felt lucky to have seen that many kills and especially privileged to have seen 06 carry out this spectacular hunt in such a determined and fearless manner.

13

The Revenge of a Mother Wolf

I N EARLY AUGUST, the main Mollie's group was still based to the south in Pelican Valley. Former Blacktail males 777 and Puff were starting a new pack with four Mollie's females, including wolf 822, who had been with the group that invaded 06's den in late April.

We were having regular sightings of the four Lamar pups up in the den forest. There were a lot of grasshoppers that year and the pups tried different techniques to catch them. They alternated between snapping at them as they jumped around and trying to catch stationary ones by pouncing on them with their front paws.

On August 3, we found 777's group on an old bison carcass a few miles west of the Lamar den. Eight of the nine Lamar adults were heading directly for them. 06 led and all the Lamar wolves had their tails raised, indicating they were in an aggressive, potentially attacking mode. I guessed that they had heard howling from that section of the valley and knew other wolves were in their territory. The group at the carcass suddenly ran off to the west with the Lamars in hot pursuit.

Female 822 split off and ran north. The Lamar pack turned away from the main group to chase her. 06 would have earlier sniffed along the scent trail of the other wolves and likely identified 822 as being one of the group of Mollie's wolves that had tried to kill her pups. The attempted den attack had been more than three months ago, and this was the first chance 06 had to strike back. I thought of the saying "Revenge is a dish best served cold."

As 822 raced toward the park road, I was with a group of wolf watchers on a low hill a few feet south of the road. We were all spellbound by what we were witnessing. The Mollie's female sped over the pavement just west of us and sprinted up a steep slope to the north. She veered east and we lost her in a series of rolling hills.

06 ran up on the road with 755 right behind her. She continued uphill on the trail of 822, but 755 turned and ran down the paved road to the east. All the other Lamar wolves ran along 822's scent trail through the hills. 755 then took a shortcut uphill that brought him directly to where 822 was running.

As I thought about it later, I figured when he reached the road he quickly took in the surrounding terrain, guessed the route 822 would be running along, visualized a shortcut, and ran that direction. I felt he was using the same observational skills he had developed while hunting elk to intercept the wolf that had tried to attack his family. 755 was soon right behind 822 and gaining. He lunged forward and pulled her down. The other Lamars ran in and joined him in attacking the Mollie's wolf.

We then heard howling to the south from the other wolves in 822's group. The Lamars looked that way and soon all but 06 left 822 and ran in that direction. 06 repeatedly bit into the other female and shook her head back and forth as she bit down. I had never seen 06 in such an intense, ferocious state. I had the thought: This is what a mother wolf does to an enemy that tried to kill her pups. 06 seemed to be channeling the fierceness and violence that characterized her grandmother, Druid alpha female 40.

There was more howling to the south. 06 paused, looked that way, then turned to 822, gave her one last bite, and walked off. I studied the Mollie's female through my scope and saw movement. 06 apparently had not given her a fatal bite to the throat, despite having had an enemy at her mercy. That reminded me of how her grandfather, Druid patriarch 21, had never killed a defeated rival. 06 was an intriguing combination of those two grandparents, who had such vastly different personalities.

I looked south and saw 06 leading her pack toward the rest of 777's group. The Lamars stopped to sniff where those wolves had been at the bison carcass. I turned my scope on 822 and saw that she was breathing. When she tried to get up, she fell, and rolled downhill. She went through that sequence a few more times.

Then I checked on 06 and saw her marking over places where the other wolves had done scent marks. That was her way of showing them this was her territory, not theirs. After that, the Lamars ran west and we got glimpses of them chasing an uncollared gray, who must have been another Mollie's female in 777's group. Doug McLaughlin saw 755 catch and pull the wolf down. She jumped up and got away.

I later followed 06 when she left her pack and headed back toward her den. She crossed the road and I last saw her carrying a stick into the den forest. I pictured her giving the stick to the pups as a toy. Just a short time earlier, she had been a fearsome avenging mother. Now she was about to return to her pups and play with them. She could be fierce when necessary, but I had predominantly seen her act in caring and cooperative ways, just like her great-aunt 42, the wolf who had raised her mother.

I went back to the attack area and got a normal signal from 822 at 12:55 p.m., six hours after 06 had left her. The transmission was beating at twice the normal rate two hours later. That was the collar's mortality signal.

Four of us from the Wolf Project—Kira Cassidy, Erin Stahler, Hans Martin, and I—walked to 822's body and examined her. There were wolf bites all over her hide. There were too many bites and too much blood loss for her to have survived. We solemnly carried her down to the road and put her in the back of a pickup so she could be driven to a place where a full necropsy would be done. A large crowd was nearby, watching us. I called them over so they could have a few moments with 822. I then spoke about the history between the Mollie's and 06's family, including the recent attempted den raid.

After that incident, Blacktail males 777 and Puff continued to travel with a contingent of young Mollie's females.

They mostly stayed in the valley known as Little America, west of Slough Creek. That kept them apart from the Lamar Canyon wolves.

MOLLIE'S ALPHA FEMALE 686 and other wolves from the main pack were in Little America on August 10. When they joined their relatives who had stayed there, the combined count was thirteen. In the evening, they were seen chasing 777, the Blacktail male. When I checked the next morning, I saw the Mollie's wolves but there was no sign of 777, despite the fact that I was getting a loud signal from his collar. Then I got a call that a dog had been staring at a spot and barking. I walked out there and found 777's body underneath some sagebrush. The many wolf bites on his body made it easy to determine the cause of death. 777 was one more wolf descended from the Druids killed by the Mollie's. We now knew of eight wolves in the north the Mollie's had killed, in addition to the Mary Mountain alpha male to the south.

777's death ended that particular round of violence between the Mollie's clan and the Lamars and other wolves descended from the Druids. 686 soon headed back to Pelican Valley with the main pack. The Mollie's had mostly been in the north for eight months by that time. The females who had joined up with Puff and 777 stayed behind.

After the departure of 686, I thought about why she and her allies had killed male 777. Since she was still without a mate and never had any pups, was she jealous of how the younger females in the family had found two males and that one of those females now had three pups? Or did she have a controlling personality that could not accept the splitting off by those independent females? Whatever was going in her mind, I will never know.

The next time I saw 06, I recalled all that she had been through during the many months the Mollie's were nearby. She and her family were greatly outnumbered by that aggressive pack, but every wolf in her pack survived that time of conflict and invasion. It was a testament to her intelligence and leadership skills.

The new group 777 had founded with Puff and the young Mollie's females became well established in the Slough Creek region. Since there was a previous pack named Slough Creek, they were named the Junction Butte pack, after a nearby landmark, and Puff became their alpha male. An uncollared Mollie's gray was the alpha female. The merging of young wolves from the two feuding families created what became a successful, long-lasting pack that continues to this day, many years later.

The Canyon White Female

It is far more common for a young male wolf like Puff and 777 to risk leaving his family to search for a mate than for a female to do the same, but a violent altercation can change that pattern. I recalled a series of events when an earlier generation of Mollie's wolves caused a young female to take on that risk and form a partnership with a former enemy.

In 2003 a light gray female and a gray male claimed Hayden Valley, a large region of open country in the center of Yellowstone, as their territory. Elk are plentiful there in the summer and fall but migrate out in the winter, making it a hard place for wolves to survive year-round. In addition, the pair found themselves squeezed between two much larger, well-established wolf dynasties: the Mollie's and the Nez Perce packs. The new group became known as the Hayden Valley pack, and the mother wolf's coat eventually turned white.

By 2007 the family had two adult daughters and they helped their parents raise a litter of pups that spring. In late October, the Mollie's wolves attacked the much smaller pack, killing the alpha pair. After that, the family's five pups were seen traveling with only one of the adult sisters. She eventually took the pups west of the park, found a mate, and established a territory there. The other sister stayed in Hayden, where she lived as a lone wolf despite the constant threat of another attack by the Mollie's.

That winter the daughter who had stayed behind paired off with a black male from the Mollie's pack. The new group settled into her parents' territory and were named the Canyon pack, for they lived near the Grand Canyon of the Yellowstone. The pair raised many litters of pups, and as the alpha female aged her coat turned white, just as her mother's had.

The story of that young female forming the Canyon pack, along with the story of the formation of the Junction pack by several young Mollie's females, is another example of wolf daughters who took on great risks and succeeded against tremendous odds. As of the writing of this book, the descendants of the dispersing Mollie's females who formed the Junction pack still control that territory, and the daughter of the white Canyon female leads the pack that holds the territory her mother claimed in Hayden Valley.

The Druid Peak pack's longtime alpha pair, wolf 42 (in the lead) and wolf 21. 21 was the 06 Female's grandfather and 42 was her great-aunt. **Kim Kaiser**

Agate Creek alpha pair 472 (howling) and 113 were the parents of 06, who was born in 2006. 472 was sired by wolf 21. **Dan and Cindy Hartman**

06 grew up to be a beautiful, fiercely independent wolf who was courted by many male wolves but rejected all of them for many years. **Robert L. Weselmann**

In 2009, when 06 was middle-aged, she picked yearlings 754 (right) and 755 (middle) to help her form the Lamar Canyon pack. She had pups the following spring. **Jimmy Jones**

06 was fearless when it came to bears. Here, she and 754 team up to drive a grizzly off a kill her pack had made. Later she brought meat back to her pups.
Ray Laible

A superior hunter, 06 was famous for making solo kills. Here she is about to attack an elk three times her size. After an epic battle, 06 won the fight.
Jeremy SunderRaj

Mollie's alpha female 686 feeding on a bison calf. 686 killed many wolves that lived in rival packs. **Peter Murray**

The Mollie's assessing the Blacktails. The third gray from the right is 686. Minutes later they caught and killed a big male who had been a suitor of 06. **Peter Murray**

686 mounted an assault on 06's den just a few days after 06 had her pups
In this photo, 06 is leading a counterattack on a group of Mollie's wolves.
Joe Allen

06 leading her family over snow on a peaceful day in Lamar Valley, with 754
to her left and alpha male 755 close behind. **Eilish Palmer**

06 and 754 crossing the road in Lamar Valley as I hold a Stop sign to keep cars from blocking their travel route. **Eilish Palmer**

When a herd of bison threatened to trample 06 while she was recovering from tranquilizer drugs after being radio-collared, Wolf Project biologists Colby Anton and Rebecca Raymond carried her to safety. **NPS/Julie Tasch**

Two of 06's pups from her first litter. During her life, all her pups survived their first calendar year. 06 was a supremely successful mother. **Peter Murray**

06's daughters 776 (right) and 926 (left). Both grew up to be alpha females.
Eilish Palmer

926 during her tenure as the Lamar Canyon alpha female. **Kira Cassidy**

926 with her five pups. One would grow up to be the pack's next alpha female. At least fifty of 06's ancestors, relatives, and descendants became alpha females. She was part of a royal dynasty. **Peter Murray**

14

The Lamar Pack
in Late Summer

IN AUGUST 2012, I spent an evening watching 06's four
pups playing in a meadow. The two black pups sparred
with each other and played chasing games. I turned my
scope to the two gray pups and saw them romping around
and chasing each other. Then all four came together and trotted around the meadow as a unit.

On the eighteenth of that month, the four Lamar pups
followed the adult wolves out of the den forest to the west,
the first time they had traveled that far. The pack was headed
for a bison carcass, where they would feed for many days.
754 was acting like a playful yearling. He went to one of the
young females and sparred jaw to jaw with her, then chased
her around. After that, he wrestled with Middle Gray. 755,
who was the same age as his brother, was bedded down
nearby and not involved with the play session.

The next morning, I saw 06 in the middle of a play group. She nipped at a yearling, ran off in a romping gait, then came back and nipped at the same yearling. I wondered if 06 was in such a carefree mood because she sensed that the Mollie's pack had left her territory.

A few days later, most of the adult Lamars crossed the road to the south to feed on another bison carcass. This time, 754 stayed behind with the pups. It looked like the four pups wanted to join the main group, but 754 led them uphill, away from the dangers of passing cars and the large crowd of people. He continued to impress me, for in this case it seemed that he had passed up going to the carcass for a meal so he could watch over the pups.

In early September, 06 and the other eight Lamar adults left the pups up at the den and headed down to the road, likely to go on a hunt. A large bus and many cars stopped exactly when 06 wanted to cross. A law enforcement ranger and I cleared a section of the road of vehicles, then stopped all oncoming cars. 06 looked down, saw that opening, and led her family through. All we had to do was create a break in the traffic, then let 06 do the rest.

WE FIRST SAW 06's pups south of the road in mid-September with seven of the adults. The pups soon found the nearby Norris rendezvous site. Many generations of Druid pups had liked to hang out there and these pups seemed drawn to the site as well. They sniffed around, then played together.

Bob Wayne, a professor at the University of California, Los Angeles, was in the park at that time. He is considered the world expert on wolf genetics. I asked about the genetics of the original wolves in the Rocky Mountains, compared with the Canadian wolves brought in for the Yellowstone reintroduction. He referred me to a 2005 genetic analysis of samples from twenty-two wolves collected prior to 1917 in western states such as Wyoming, Utah, Colorado, and North Dakota, along with samples from several plains states and New Mexico. The study, with Jennifer Leonard as the lead author, discovered that those twenty-two wolves were very similar genetically to wolves currently living in the western parts of the United States and Canada, which means that the wolves reintroduced to Yellowstone in the mid-1990s from western Canadian populations were essentially the same genetically as historical wolf populations in the American West.

The paper also addressed wolf dispersal and migration in North America after the last ice age. Glaciation reached its maximum extent about eighteen thousand years ago and covered most of the Canadian Rockies, meaning local populations of wolves and other species there were wiped out. But during that time parts of the American Rockies were free of glaciers, and wolf populations survived in those regions, known as refugia. Those gray wolves, in the words of the research paper, "served as a source of gray wolf colonists for deglaciated Canada."

That last quote means that wolves from what are now the northern Rocky Mountain states gradually dispersed north into newly ice-free areas and eventually repopulated regions in Canada, where the original wolves had died off when glaciers covered the region. That was a prehistoric wolf reintroduction that operated in a reverse direction to

Yellowstone's reintroduction, when wild wolves were captured in Alberta and British Columbia and released in the park. Another way to sum up all that information would be to say wolves from the American West recolonized western Canada after the last ice age, and in modern times wolves from that part of Canada were brought south and recolonized the northern Rockies.

Bob recently sent me a 2016 paper on genetic subdivision in wolves, with Rena Schweizer as lead author. My understanding of this research is that all North American wolves are the same species, but that there are distinct populations in six habitats with differing climates, vegetation, and prey animals. The wolves in those habitats have learned to adapt to local conditions and pass down their survival knowledge to their young. These populations within the gray wolf species are known as ecotypes, with such names as Arctic, Boreal Forest, British Columbia, and Atlantic Forest. In my correspondence with him, Bob wrote, "The Yellowstone wolves, current and historic, were almost certainly part of the Rocky Mountain ecotypes and continuous with those regions in Alberta and interior British Columbia which share genetic similarity to the western (United States) wolves in the 2005 study."

As I thought about those findings, I realized that the thirty-one wolves brought to Yellowstone from Canada would live the rest of their lives in a region where their distant ancestors likely roamed thousands of years ago.

I then realized how my genetic background has some similarity to the wolves researched in those two papers. My father's ancestors left the Scottish Highlands around 1800, during a time of forced removal known as the Highland Clearances. They emigrated to Canada, and later descendants ended up in Massachusetts, where I was born. And so my DNA should be very similar to that of the Highland people whose ancestors never left the homeland.

A WEEK AFTER the Lamar wolves arrived at the Norris rendezvous site, the adults took the four pups across the Lamar River to Chalcedony Creek, where the pups immediately began to play in the meadow. A few days later, the pups followed the older wolves west.

The alpha pair were especially playful on that trip. 755 chased 06. She easily outran him but turned around and raced back. Other adults had to dodge her or risk getting knocked over. 754 jumped out of her way, then chased her. She also outran him. 06 came back to 755 and got him to chase her. It looked like she was showing off how fast she was. The younger adults were in their own play group. 06 ran over, jumped up on some of them, then ran off. After that, she ran circles around other wolves.

Then a black pup invented a new game. It found a plastic bottle, picked it up, and ran around the other pups, like he was daring them to chase him. They pursued the black and when that pup dropped the bottle, another pup grabbed it and ran off. 06 and other adults saw the game and raced over. Soon a gray yearling had the bottle. 754 chased her and when she dropped the bottle, he snatched it up. As he ran through the main group of wolves, the bottle slipped out of his mouth, so he ran back and grabbed it once more. Then he put it down and a yearling ran off with it. That wolf soon let go of the bottle and a gray wolf got it next. All that chasing with the bottle looked a lot like a rugby game. I always thought that wolves would have no understanding of human sports with two exceptions: rugby and wrestling.

When that game got boring, the wolves continued to play in other ways. 755 and 06 came together and sparred with their jaws. Then he suddenly ran off. She did not chase him, so he sped back and ran in circles around the wolves. He sparred with 06 once more, then resumed running in circles. As he did that, 754 played with 06. Seeing all that hectic activity in a pack of wolves was no different than watching a group of neighborhood dogs playing with each other.

THE ADULT LAMAR wolves were in Silver Gate in late September. At one point they were visible from the paved road and many people stopped to watch them. That location was a legal Montana wolf-hunting zone, and two local men in the group had licenses to shoot a wolf. Hunting regulations prohibit shooting at an animal from the road but sometimes that restriction is ignored. One of the men, who had a local business, later told a friend that he did not shoot a Lamar wolf because he made a lot of money from wolf watchers and did not want to risk a boycott of his store.

A few days later, we heard the Lamar wolves howling from near the park border. I then heard a man play his bagpipes from his nearby cabin. He paused, probably to see if the wolves would howl in response to his music. He tried eight times, but the wolves never did howl back. I remembered that there was a famous blues singer who called himself Howlin' Wolf. Maybe if we had brought him to Silver Gate, the wolves would have howled back to the sound of his voice.

Wolf Project biologist Matt Metz downloaded data from 06's collar and got the GPS locations where she had been. One day she was several miles east of Silver Gate, over the state line in Wyoming. At that time, there was a quota of eight wolves that could be shot in that section of the state. Apparently, no one with a wolf-hunting license saw her, for she was safely back to her den by 1:00 a.m., about twenty-two miles from her previous GPS location.

Late in the evening of October 20, I heard eight gunshots in Silver Gate. I had gotten signals from 06's family that day in Lamar Valley, so I knew they were not near our town. But the Lamars returned to Silver Gate a week later. The pack was back in the park the next day and all the members were accounted for. I was getting worried by the pack's movements in and out of the park into a zone where it was legal to shoot them.

EVERY FALL NATIVE Americans John Potter and Scott Frazier came to the park to conduct a wolf blessing ceremony. They always invited people to join them. Their initial blessing took place in early 1995, when the first fourteen wolves from Alberta arrived in the park. During the 2012 ceremony, Scott made a profound statement about the success of the park's wolf reintroduction program: "It is good to be part of putting something back rather than taking something away." That was a perfect way to describe the success of the program.

Then John sang a traditional Native song and led the blessing, which was done as a group. That involved the burning of *wiingashk* (sweetgrass). After that, he said something that deeply moved all of us: "The original Canadian wolves that were brought in on the day Scott and I did our first blessing are gone now, but their spirits live on in the surrounding mountains."

I would like to think that the spirits of later Yellowstone wolves who were no longer with us were also up in those mountains, wolves such as 21, 42, 302, and 480. While working on this book, I asked John about that, and this is what he wrote me: "Yes, absolutely. In fact, I would say that the Spirits of ALL the Park Wolves—even those who lived in the Park prior to the extermination in 1926, and all of their ancestors, still live and move and thrive in the mountains here. I think the Spirits of the Wolves who lived here before 1926 actually CALLED those Wolves who returned in '95 and '96. They called through you, Mike [Phillips, the first leader of the Wolf Project], Doug [Smith], me, Scott—through everyone who was even remotely involved in bringing them back."

Of all the people who have ever lived on Earth, I feel the Indigenous People of North America know and understand wolves the best, so I put a lot of stock in what John told me. I later read in *Yellowstone Wolves* that wolves had lived in what is now the park area as far back as fifteen thousand years ago, so based on what John said, there were a lot of wolf spirits calling for their kin to repopulate Yellowstone.

BY THE FALL of 2012, the balance of power between the Mollie's and o6's family had drastically changed. There were now only six wolves in the Mollie's pack and thirteen wolves in o6's family. That was partly due to the departure of the Mollie's females who started the Junction pack, and to the fact that those wolves were not acting aggressively toward the Lamar pack. 06 had successfully steered her family through a period of dire threats, but she and her family would soon have to deal with a much different deadly challenge.

15

Hard Times

I N EARLY NOVEMBER, most of the Junction wolves were on a bison carcass north of the road at the west end of Lamar Valley. 06's pack was south of the road, traveling west. When the Junctions howled, 06 turned around and led her family away from the other wolves. I interpreted her actions to mean that she chose to avoid an unnecessary fight with another pack that so far had not bothered her family.

The Lamars were soon south of the Yellowstone Institute. The law enforcement rangers had a shooting range nearby, and we could hear them firing their weapons. The Lamar wolves repeatedly looked toward the sound of the shooting and rushed off away from the gunfire. That was a good thing to see. We wanted the wolves to be wary of the sound of guns when they were outside the park during the hunting season.

On the morning of November 11, I got a message from a friend in Wyoming who researched wolves there. He reported seeing the Lamars out of the park in a wolf-hunting zone. The next morning, I got another text from him. It said he had just seen twelve of the thirteen Lamar wolves up high on a ridge. 754 was missing and he was not getting his signal. When I got home that evening, I called the Wyoming Game and Fish wolf hotline and heard that a wolf had been shot earlier in the day in that zone. I immediately thought that it could be 754. There was a limit of eight wolves that could be taken in that zone. This was the seventh wolf shot.

The following day, November 13, we got confirmation that 754 had been killed twelve miles east of the park. I called my friend in Wyoming and he told me the Lamar wolves were heading back into the park. He added that he had spoken with the people staffing the local game-check station. They told him that a hunter had stopped to report shooting a wolf. The staff checked the animal and the man's hunting permit. The radio collar identified the wolf as 754. During the conversation, the hunter told them he wanted to shoot the biggest wolf he could find. 754 was the biggest member of the Lamar Canyon pack.

Later that day, I got a mortality signal from wolf 823, one of the Mollie's females who had helped start the Junction pack. It was coming from just north of the park border in Montana, to the east of Yellowstone's North Entrance. We soon got confirmation that 823 had been reported as a kill by a hunter. A lot of elk had been shot in that area and the gut piles left by hunters were a big attraction for ravens, coyotes, and wolves. The Junction wolves were now down to seven adults and three pups.

At the time, that Montana hunting zone did not have a limit on how many wolves could be shot. The killing of

Junction female 823 started a movement to request that the Montana Fish and Wildlife Commission impose a quota, and the proposal was eventually approved. The public reaction to her death saved the lives of many other wolves in the coming years. If she had not been collared and known to be a Yellowstone wolf often seen by park visitors, that limit on shooting wolves across the park border might never have happened.

We found out that 823 had been shot by a man who worked for a local hunting outfitter. His boss had told him not to shoot a collared wolf, but he killed 823 despite that order. The boss fired him.

DOUG DID A flight on November 15, three days after the death of 754, and called down to tell me he was getting the other Lamar signals in the park. They were in a forest east of Round Prairie, but he could not get a count. Later that morning, I saw 755 leading the pack toward their den forest. I got a count of twelve. That confirmed that 754 was the only member lost during their trip into the wolf-hunting zone.

06 led when they entered the den forest and later was still leading when they reappeared farther west. She frequently stopped and did a lot of howling as she looked around. The rest of the pack joined in. When they paused, she continued to howl by herself. Soon the others resumed howling as well. It occurred to me that the Lamar wolves did not know what had happened to 754. All they knew was that he was missing. It made sense for them to come back to their home base and look for him there. Howling was the easiest way for them to try to find him. But I knew they were calling out in vain. 754 was never going to howl again.

754 lived the life of a wild wolf and never bothered any people or livestock. He was a big, strong male who seemed to have no ambition to be an alpha. 754 appeared to be content to help his family hunt and care for each new litter of pups. He especially liked to play with the pups and young adults. 754 was a tough, courageous wolf who stood up to sixteen rival wolves and could win a confrontation with an angry grizzly bear, but had no defense against a high-velocity bullet.

Many of my friends have told me of how their dogs will go through a period of depression and grieving when a canine companion dies or goes missing. I think the Lamar wolves went through the same type of grieving process for their family member and friend.

Even if they were grieving, they still had to eat. On November 21, they scavenged on an old bison carcass. I saw a pup carry off scraps. I checked my records and found that the bison had died over 113 days earlier.

Early on December 3, I got signals from the Lamar wolves as I was leaving Silver Gate. I wondered if they had traveled here during the night to look for 754. Later I saw them heading back west to Lamar Valley. The next day, I found them near their den and the twelve wolves did a lot of howling. I felt they were still trying to contact 754.

The pack then went south, up toward the crest of Mount Norris. I lost 06 and the others over the top at 10:10 a.m. I did not know it at that time, but they were heading back to where they had last seen 754. Thinking back to that day, I believe the Lamar wolves were going there to search for their missing family member. I knew they were risking their lives, for one more wolf could be shot in that hunting zone outside the park.

I DID NOT get any Lamar signals the following two days, December 5 and 6. I did my last check for them at 4:40 p.m. on that second day, then headed back to Silver Gate. At 7:12 p.m. Dan Stahler called my home number and told me Wyoming Game and Fish had contacted the park to tell us that a collared gray wolf had been shot near where 754 was killed. Dan and I both knew that could mean it was 06. But Wyoming had a collaring program, and the wolf might be from a different pack. That is what I told myself.

I left my cabin at 6:44 the next morning, drove east a few miles to check on signals, then turned around and headed out into the park. I got no signals from the Lamar wolves anywhere. At 7:25 a.m. I received a text message and pulled over to read it. I saw that it was from Dan and had a good idea what it was going to say. He had gotten confirmation that the shot wolf was 06.

I drove on and heard that the Junction wolves were visible at Hellroaring. It would take me about a half hour to get to that location, and I knew that a large crowd of wolf watchers would be gathered there. I spent that time trying to plan out what I was going to say to all the people who were waiting to hear what we knew about the wolf who had just been shot and were desperately hoping it was not 06.

I got to the lot, saw some of the Junction wolves, then turned to everyone and told them the news. Each person looked at me in stunned silence, then the emotions of the moment hit them. One woman collapsed and started sobbing. I knelt down and did what I could to comfort her. Then I went around to everyone else and we shared stories about 06 and 754. That helped a bit.

We watched eight of the Junction wolves and that worked to take our mind off 06's death. Then four Blacktail wolves walked into sight, including 06's sister, alpha female 693, and alpha male 778, the former Druid wolf. They came together and played for some time. Seeing all those wolves began to bring me out of the sadness at the loss of 06. When I got home that evening, there were a lot of messages on my answering machine. I felt it was my responsibility to call all those people back and to contact others who knew 06 well. Doug Smith called to console me. I told him about the many wolves we had just seen and how they were living out their lives in a normal way. Despite what had just happened outside the park, my time with the Junction and Blacktail wolves that day allowed me to finish the call by saying, "It was a good day in Yellowstone."

06 was the eighth wolf taken in that Wyoming zone, and that filled the quota. Her family lingered there for the next few days, but her death meant that no other wolves could be legally killed in that valley. They were safe because of her.

THE NEXT MORNING, December 8, we again saw the Junction and Blacktail wolves and watched wolves the following day. I realized that it was much easier for those of us who were in Yellowstone, seeing wolves every day, than it was for people living far from the park. They were thinking about the loss of 06, as we were, but getting to watch wolves was a positive experience that helped us to balance out what had happened.

That day the *New York Times* printed an obituary for 06. It read in part: "Yellowstone National Park's best-known wolf, beloved by many tourists and valued by scientists who tracked her movements, was shot and killed on Thursday outside the park's boundaries, Wyoming wildlife officials reported. The wolf ... was the alpha female of the park's highly visible Lamar Canyon pack and had become so well known that some wildlife watchers referred to her as a 'rock star.' The animal had been a tourist favorite for most of the past six years."

On December 10, we heard that the Montana Fish and Wildlife Commission had had a meeting and voted to make an emergency closure of the zones north of the park to wolf hunting and trapping, because of the many Yellowstone wolves who had been killed there. We were grateful for that news.

Doug McLaughlin had gone to that meeting to testify and told me that the deaths of 754 and 06 had become such a big story throughout the world that the commission was flooded with demands to close that wolf-hunting zone. The two Lamar wolves were shot in Wyoming, but their deaths were the catalyst for the closure in Montana.

I later found out that Doug Smith had given the commission radio-tracking data on where Yellowstone wolves had gone when they left the park border. They used that information to draw the closure zone. That demonstrated the value of our collaring program, for it saved the lives of many park wolves.

BY THAT TIME, it had been four days since 06 was shot and we did not know where the surviving members of her pack were. In midafternoon, I got a radio call from my colleague Hans Martin. He was getting signals from 755 and 820 in Lamar Valley. Hans saw one wolf at a new deer carcass near the den forest, then got a report of additional wolves close to there. I joined Hans and saw 755 and six other Lamar wolves.

The word spread fast about the sighting, both inside the park and everywhere else through social media. The news brought a lot of comfort and relief to people who knew the Lamar wolves. By the next morning, we had accounted for eleven Lamar wolves. That meant that 755, all the younger adults, and the four pups had survived the trip back home.

As they had done when they came back after the loss of 754, the Lamar wolves did a lot of howling. I saw 755 start one of the group howls. The other wolves came to him and they all socialized together. It was a bonding ceremony that must have helped them cope with the loss of their two family members. The wolves moved on, but frequently had more howling sessions. I think that wolves express their emotion when they howl, and those vocalizations sounded mournful to my ears. I thought all that howling could be considered the wolf version of the type of music we call the blues.

I called my friend in Wyoming and we talked about 754 and 06. He told me there were a lot of elk in the open areas where they were shot. That helped explain why the pack had traveled out there. He also said that the local Wyoming Game and Fish biologist was talking about lowering the number of wolves that could be taken in that hunting zone for the next year. So, like in Montana, the deaths of the two Lamars were going to save the lives of future wolves.

In mid-December, I saw Mollie's alpha female 686 alone in Lamar Valley. She was scavenging on the old bison carcass from last August. I later saw two other wolves from her pack near the carcass. 686 mostly stayed apart from them. The once-mighty pack was at a low ebb. In all her travels into Lamar Valley, 686 still had not found an unrelated male to join her pack. Meanwhile, I noticed the Lamars were spending a lot of time near their den forest and I wondered if that gave them a sense of security and normality.

ON DECEMBER 18, the Lamar wolves went back to where 06 had been shot. They were safe there now that wolf-hunting season had been closed. I felt they had gone back to look for her and 754 once more. They stayed east of park for a few days.

At the end of the year, Matt Metz tabulated the data from 06's GPS collar. During the 307 days she had that collar, 06 was in the park or along the border in the Silver Gate region for 297 days and had been out of Yellowstone to the east on a total of ten days. She was shot on one of those ten days.

I eventually got more information about the details of her death. When the man who shot 06 stopped at the local game-check station, he told the staff that he had seen 755 and could have shot him but had noticed a bigger gray wolf near him. He had aimed at that one instead. 06 was an average weight for a female, but 755 looked somewhat small for an alpha male.

Later I spoke with a friend who lived east of the park and knew the person who had killed 06. The man told my friend that after he shot her, 755 and the rest of the pack howled from a position close to her. The shooter decided to give them some time with o6, so he left and came back an hour later. On returning he saw that the pack was still nearby. They likely did not comprehend she was dead. The hunter was required to take o6's body to the nearby check station, so he had to retrieve her. 755 and the other wolves moved off when the man and a companion went to get her. He said Lamars stayed near that site for several more days and howled continually during that time, trying to contact her.

As I thought about that story, I realized it demonstrated the bond between the two wolves. 755 must have had some understanding that this was a dangerous place, yet he loyally stayed there by the side of o6, likely licking her gunshot wound and trying to get her to stand up and leave with him. After her body was taken away, he remained close to where she had been killed for another thirty-six hours as he howled, presumably attempting to contact her. That howling would have given away his location to people who could have shot him if wolf hunting had not closed, but he did not know that. Heedless of the danger to himself, he must have stayed there until he realized that she was gone, and perhaps sensed that they would never be together again.

Writing about her death made me wonder if o6 had a few moments after she was shot to look around and see that 755 and all her sons and daughters were all right. I hoped she did that and was comforted by the sight as she slipped away.

At the end of 2012, we documented that all four of o6's new pups were still alive. She had had five pups in 2011, and four when she was a first-time mother in 2010. All thirteen had survived through the end of their first calendar year. The

Wolf Project keeps track of pup mortality and tabulated that the average survival rate by the end of the year is 73 percent. The rate for 06's three litters was 100 percent, an indication of how good she had been at running her pack and raising the family's pups.

In 2014 the National Geographic Wild channel aired a documentary on 06 that was made by local filmmaker Bob Landis. It was titled *She Wolf*, and millions of people saw it all around the world. Bob later interviewed me about 06 as an extra feature for the video release of the program. I also did an interview on 06's life story for the NPR radio show *Snap Judgment*. They spoke with Doug Smith as well. The episode aired on May 23, 2014, and became one of their most requested rebroadcasts. It was originally titled "Legendary," but when they repeated the episode, they renamed it "The 06 Female."

PART IV

———◆———

2013

16

Starting Over

WHEN AN ALPHA female as accomplished as 06 dies, it has a devastating impact on her mate and family. I felt my role in her story was to document what would happen to 755, her adult sons and daughters, and her final litter of pups.

At the beginning of 2013, 755 and his daughters 776 and 820 were still in Lamar Valley but the other family members remained outside of the park. The February mating season was approaching and the older females east of the park were probably looking for mates in that section of Wyoming. Soon several males from the disbanded Hoodoo pack joined those Lamar wolves. The group was eventually named after their lineage, so they became the new version of the Hoodoo pack.

755 was too closely related to 776 and 820 to breed with them. I expected that he would have to set off on his own to seek out a new partner, and indeed he soon broke away from his family. He passed through several wolf territories, and I

saw him sniffing at wolf scent trails in each one. I figured he was looking for the scent of a single female. In late January, I saw 755 with a gray Mollie's female in Lamar Valley. The next morning, they flirted face to face. The female, 759, was especially affectionate with 755. She got in front of him and put both front paws on his shoulders.

Doug Smith was doing some collaring soon after that and had the helicopter pilot fly to Lamar Valley so he could dart 755 and put a new radio collar on him. But as the pilot got close enough for a shot, 755 ran under the helicopter. That put him in a position where Doug was blocked from getting a clear shot. 755 did that repeatedly and the crew had to give up on him. I later heard that when 755 was east of the park, Wyoming Game and Fish had tried to capture him with a netting operation and he apparently learned that they could not get him if he got under the helicopter. I took that as a sign of how intelligent a wild wolf can be.

I continued to see 755 and the Mollie's female 759 together and was glad that he had found a new mate. One day I saw him feeding on a bighorn sheep he had killed near the road. 759 had grown up in a remote part of the park and was not used to traffic, so she stayed well away from the sheep and the road corridor. The two wolves howled back and forth. After stuffing himself, 755 went to her and regurgitated a big pile of fresh meat for 759 to eat. Later I saw them bedded down side by side, chewing on the same bone. All that indicated 755 had abdicated his alpha male position in the Lamar pack to be with her. He had moved on and was starting a new chapter in his life.

A FEW DAYS later the main group of Lamar wolves came back to Lamar Valley. They howled from the den forest. I did a check and got good signals from 755 and 759 just to the west. After that, I saw nine wolves chasing 759, then lost sight of all of them. 755 was howling continuously nearby. He must have been trying to contact her. Soon after that, I found them together as 759 followed 755 through the Chalcedony Creek rendezvous site. She had blood on her coat. When she stopped, 755 went to her and licked her wounds. Then both of them went into a forest. Two days later, I got a mortality signal from her collar in the direction of those trees. 755 left the valley and went west, alone once more. Doug Smith and others examined 759 and found many wolf bites on her body. She was pregnant and her unborn pups were likely sired by 755.

Since I knew 755 well, I felt so sorry for him. He had lost his brother and 06 a few months earlier, found a new female to pair off with, and now had lost her as well, along with a litter of pups. She had had the bad luck to run into the remnant of the Lamar pack. 759 was a Mollie's wolf and the Lamar wolves likely would know that from her scent. They would have had a reason to go after her because of the den raid incident.

The 2013 breeding and denning season now seemed to be a lost cause for 755. He would have to wait another year for a chance to have pups. I wondered how he could cope with so many losses in his life. Then I thought of myself. My father died suddenly when I was ten years old. Many years later, when I was working in Alaska at Denali National Park, a friend, wolf biologist Gordon Haber, died in a plane crash while researching wolves. Another Alaskan friend committed suicide and a third was killed by a grizzly bear. People experience tragedies and losses and keep on living their lives, and I felt 755 would do the same.

ON APRIL 3, I saw 755 with a gray female near Blacktail Plateau. She may have been from the Leopold or Eight Mile pack. The two wolves were together on a fresh kill the following day and they played a lot together. A few days after that, I heard that 755 had rolled down a snow patch, then lain on his belly and slid the rest of the way down the slope. The gray female watched him and copied his slide. They were still together on April 20 and the following day. They had not met up until after the February mating season, so they could not have pups that spring.

Unfortunately, 755's relationship with this new female did not last long. Our final sighting of the two of them together was on the north side of Lamar Valley on June 30. The pair had been near two neighboring packs recently and perhaps she did not want to risk being attacked by them. 755 would be a lone wolf in the coming months as he traveled throughout the northern part of Yellowstone.

By early September, his once-black coat was mostly gray. In poor lighting conditions he sometimes looked like a gray ghost. He had been born in 2008, so he was now over five years old, about the average life span for a Yellowstone wolf. I wondered if his days as an alpha male were over. I still hoped he might find a female as extraordinary as 06 to help him start a new family, but that seemed unlikely.

755 FAILED TO raise a family in 2013, but two of his daughters were more successful. The main group of Lamar and Hoodoo wolves had settled down east of the park and denned there. But female Middle Gray stayed behind in the original Lamar territory, making her the lone representative of 06's family still in the pack's territory. She made kills on her own, a sign she had learned well from her mother. She was later joined by her younger sister, black female 926. I wondered if these two females would be able to revive the Lamar pack, but to do that they needed an alpha male.

A gray uncollared male with an injured ear that stuck out at an odd angle crossed the road one day and went up to the den forest where the two Lamar sisters were often seen. He was unknown to us but seemed to be a dispersing male looking for a mate. The two sisters were looking for just such a wolf, so he was heading to the right place. The three wolves formed a new version of the Lamar pack. Later that male was collared and given the number 925. We eventually learned that he originally came from the Hoodoo pack.

Middle Gray looked very pregnant on April 18 and likely was about to have pups at the den forest. Her sister 926 and the new alpha male 925 frequently left the den forest together to go on hunts or visit carcasses. I often saw them playing. He would do a scent mark and she would mark over it. That normally was a sign that two wolves were the alpha pair, but I had many earlier sightings of 926 acting subordinate to her older sister. I watched as 926 fed on a new carcass. She went to the den and was back at the carcass eleven minutes later. That quick round trip indicated that 926 had likely regurgitated meat to her sister at the den and returned to the carcass. She then did a ten-minute round trip from the kill to the den and back. Later in the day, she did two round trips.

Middle Gray was seen chasing elk in Lamar Valley on May 10. I saw that she had signs that she was nursing pups. That meant that she was raising 06's grandsons and granddaughters in the den forest where 06's mother had been born in 2000.

926 made a big contribution to the family when she found a big bison that had died of natural causes. Middle Gray followed her sister's scent trail and fed after 926 left. That huge carcass would keep the pack well fed for many days.

The next day, the two sisters came together. This time, Middle Gray was subordinate to 926. She was a year older than 926 and had been dominant to her. But perhaps she was just acting lower-ranking during the denning season so 926 would continue to help her with the new pups. In the past, I had seen other nursing alpha females behave the same subordinate way and felt that was because they needed a lot of help from other females in the pack. It is a smart move to treat others well when you have to depend on them.

By May 26, the big bison carcass had been totally consumed. That morning I followed Middle Gray trotting toward the remains. She stopped, sniffed around, then dug with her front paws into the dirt. Soon I saw her pull out a big piece of bison meat one of the other wolves must have cached. Middle Gray walked off and sniffed at another patch of ground farther away. Then she reached down and without doing any digging came up with something she ate. After walking away, she repeated the process and got another meal. I thought she was eating eggs out of ground nests. The next day, I saw 925 do the same thing and later witnessed 926 getting eggs twice. I had previously seen wolves find and eat eggs along Slough Creek and figured they had been laid by waterbirds such as Canada geese. These sites were well away from water, so the eggs had more likely been laid by songbirds such as sparrows, meadowlarks, and horned larks.

At the end of May, all three Lamars came together for a mutual greeting. They walked together in a tight parallel formation and I noticed that 926 was squeezed in between her older sister and the male. The scene looked like two human sisters walking with a boy they both liked, and the younger sister was making sure she got to be next to him.

There was something about that moment that brought home to me how much I liked 926. Her mother had been killed and her father had left the family, but she was carrying on with an appealing, positive attitude about life. As I thought more about it, I realized that is exactly what all the great female wolves I had known consistently did: when times of trouble and tragedy came their way, they never gave up but always carried on.

Jeremy SunderRaj, the high school student, told me that he had seen the three Lamar wolves crossing the river south of the den forest. 926 got out of the water first, then turned around and playfully tried to block Middle Gray and 925 from climbing up the riverbank. Later I saw 926 tease the male by running circles around him. As I thought about those incidents, they made me feel good. 926 was instigating play sessions, and I took that as a sign that this new version of the Lamar pack was getting back to normal.

I SAW MIDDLE Gray's pups for the first time on the evening of July 13. There were two and both were black. We know that two gray wolves cannot have black pups, so Middle Gray must have gotten pregnant by a black male when she was east of the park in the February mating season. There had been a black male in that group during that time. This meant that 925 was raising pups that had been born to another male, yet he worked hard to support them and seemed to treat them no differently than if he had been their sire.

The next day, I saw Middle Gray lick the face of 926 in a subordinate manner. I had seen her act that way to her younger sister in May but at the time figured she did it to solicit a feeding while she was stuck at the den with her newborn pups. Now it looked like 926 was the alpha female of the pack and would be the sister pairing off with 925. She would serve in that position for many years, and that meant the long-term fate of the pack was on her shoulders.

From what I have observed over the years, young female wolves consistently seek out unrelated males to mate with, a practice that we know gives their pups a healthier mixture of genetics than breeding with a relative. But there are examples such as this one where wolves like Middle Gray do not stay with that initial partner and end up with another male who had not bred her. In every case that I know of, that new male raises her pups in a manner that appears to be no different than if they were his biological sons and daughters.

It is far more common, however, for a female to choose a male, mate with him, and stay with him for life. That suggests that in addition to finding an unrelated male, she chooses a male she judges to be reliable and dependable, someone she can count on for the long run to keep the family safe and well provided for. This choice will be the most important one of her life.

The story of 755 and his daughters illustrates another aspect of wolf society. 755 had been the founding alpha male but left the pack when 06 died and took on the great risk of trespassing into the territories of rival packs to seek out a new mate, a strategy that could easily have gotten him killed. His departure made it far easier for his daughters to attract a mate. The same thing had happened when Druid alpha male 480 lost his mate and left the pack.

IN MID-JULY, WE were in the middle of the bison mating season. A big bull would find a cow who was getting ready to breed and follow her around while constantly grunting. As soon as he bred with her, the bull would take off to look for his next female. All of this contrasted with the long-term affectionate relationships wolf pairs have. Once, Laurie Lyman told me her worst nightmare would be to come back as a cow bison.

The two Lamar sisters took both of the pups on a walk to the east of the den forest on July 20. Both females chased a mule deer. That would be the first hunt witnessed by the pups.

I realized that all the sightings I was having of the Lamar wolves and their pups had taken my mind off the deaths of 06 and 754, as well as the departure of 755. Middle Gray and 926 were carrying on what those adults had started and were raising the next generation of pups, with the help of new male recruit 925.

At the end of July, we got a good look at the two Lamar pups. One chased grasshoppers as the other dug at a ground squirrel burrow. The first pup joined its sibling at the burrow and one of them went down into the hole. When it came out, I saw that the pups were different sizes. As they came from a single litter, the size difference indicated that one was likely male and the other female. A pup dropped into the ambush position, then jumped up and chased its sibling when it came by. The pair then ran around together. One pup tripped and fell but got right up and resumed running.

Both black pups were up at the den forest with Middle Gray on August 4. After that, we saw only the smaller female pup. One day I saw that little pup go to 926 and paw at her face. 926 licked the pup in response. When she paused, the pup pawed at 926's face again, and that got her to resume licking the pup. I saw this pup trotting after the two Lamar sisters on August 15. She did a good job of following them as they traveled several miles east at a good pace. The pup romped around with 926 and seemed to be in good health. I lost them heading into a forest.

Middle Gray continued to act subordinate to her sister, so we were sure that 926 was the alpha female, despite being younger. The pack continued to be based at the den forest through August 25.

The Initiation

When I saw 926 licking that pup and how it pawed at her face to get her to continue the licking, I remembered watching Agate alpha female 472, 926's grandmother, and a pup go through that exact sequence in late 2010. 472 must also have licked 06 in a similar manner when she was a pup, and I am sure 06 licked 926 the same way in her early months. That maternal behavior would have transferred from 472 to 06 to 926. I had also witnessed the adult Blacktail wolves in a ceremony where they licked each other's faces, including their alpha female licking a lower-ranking female.

I recalled Frans de Waal's comments on how mutual grooming among primates works to bond members of a group together and decided to do more research on wolf social behavior. I went through my twelve thousand pages of field notes and searched for examples of licking among wolves and found thousands of observations. Then I picked out 333 representative cases and divided them into four major categories: adults licking adults, mother wolves and other adults licking pups, pups licking adults, and pups and mother wolves licking another pack member for a regurgitation.

Some of the cases seemed to be a form of reconciliation, such as the time when a new male came into a rival pack's territory, fought with the old alpha male, and defeated him. The victor let the former alpha stay in the pack and licked the wounds he had inflicted on him. From that point on, they worked together as a team. Back in the era of wolf 21, there was a time when one of his yearling sons was picked on and bitten by a group of other young wolves. 21 went over to the victim and licked him, something that seemed to be an act of consolation. There was a peak in licking observations in the mating season, when courting males and females would lick each other in an affectionate, intimate manner.

I remembered that years ago I had visited the Canadian Centre for Wolf Research in Nova Scotia, and that its director, Jenny Ryon, had given me a videotape shot underground in a wolf den at its captive facility. I got it out and watched it. The footage began with a pregnant wolf in the den. She soon started to give birth, and as a pup came out of her, she licked it. I replayed that moment many times and concluded that the trauma of being born would be counterbalanced by the pleasurable, reassuring feeling of being licked. As the video continued, there were many other examples of the mother wolf licking her three pups as they clumsily crawled around the den and later started to learn to walk.

I saw in Yellowstone that when pups came out of the den, the father wolf and other pack members would lick them and the pups would lick them back. That mutual licking would serve to welcome and integrate the young pups into the pack. Licking remains a constant throughout wolves' lives, a family bonding ceremony repeated thousands of times.

All of this went back to the critical moment of birth and how that mother wolf licked her pup. I realized that was the pup's initiation into wolf society. If it was a female, she would eventually lick her own pups the same way her mother had licked her at birth, and her daughters would do the same thing with their pups. It would be an endlessly repeated socialization ceremony going back to the origin of wolves. This underscored for me how critical mothers are in wolf society.

After thinking about mother wolves licking newborn pups in the first moments of life outside the womb, I thought about 755 staying with 06 after she had been shot and how I felt he had likely licked the spot where the bullet had entered her body. Visualizing that made me think that for some wolves like 06, the experience of being licked at birth by her mother would be matched by being licked by her longtime mate in her last few moments of life.

17

The Setback

IN SEPTEMBER, MOLLIE'S alpha female 686 died. Wolf
Project staff examined her body and concluded that she
had been killed by other wolves. Her pack had dropped
down to just three adult members that summer: 686, adult
female 779, and a new gray male. Both females had denned
that spring, but after a few weeks 686 left. We never knew if
686 was driven out by 779, who had become the pack's alpha
female, or if she left by choice. No pups were ever seen at
her den. 686 traveled around as a lone wolf while 779 and
the gray male raised five pups back at the den. As far as we
knew, 686 never had any surviving pups during her five years
of life.

An ironic part of 686's story is that for years she searched
for an unrelated male to mate with, but when a male finally
joined the pack, it was a younger female who had surviving pups, not her. If you evaluate an alpha female wolf on
the basis of pup production and survival, 06 was supremely

successful. She had fourteen pups in three years and raised all of them to adulthood.

686 had presided over a time of great violence. During her reign, the Mollie's killed at least nine wolves. She also invaded the Lamars' den site, where she tried to attack 06 and her family. Like Druid alpha female 40, wolf 686 lived a violent life, and both eventually had a violent death. If they had been characters in the Star Wars films, you could say that they both went over to the dark side.

That year, 2013, 686's former pack made only one trip to the northern section of Yellowstone, so the long feud between the Mollie's and the Druid genetic line, including 06's family, seemed to be over.

Later the new Mollie's alpha female, 779, was recollared and weighed. She was 136.6 pounds, the biggest female ever known in Yellowstone. In contrast, when 06 was a five-year-old she was 94 pounds, close to the average weight for a park female. I recalled how fast 06 ran when chasing elk and realized that being too heavy could be a disadvantage for wolves like the Lamars who specialized in hunting elk, which flee when chased. The Mollie's, however, specialized in hunting bison, who often stand their ground, so having bigger wolves in the pack gave them an advantage in fighting such large opponents. Despite being such a formidable-looking wolf, 779 did not exhibit the kind of super-aggressive behavior toward neighboring wolf packs that 686 did.

ON SEPTEMBER 30, we heard on the news that the federal government would be shutting down at 10:00 p.m. because of a political dispute in Washington. That meant that

Yellowstone would be closed to visitors, and all government workers, including me, would be placed on furlough.

Those of us who lived in Silver Gate and the neighboring town of Cooke City could drive through the park to buy supplies, go to medical appointments, and visit friends, but we could not stop anywhere to walk around or look for wildlife. There were massive disruptions for people traveling to Yellowstone. Many of them had dreamed for years of seeing the park. When they finally arrived, they were told they could not enter due to the shutdown. It was a sad situation because the problem was created by politics.

I would go out every morning to look and listen for wolves east and west of my cabin, then turn around at the park entrance. There was a National Forest trail nearby that I often hiked to the park border. In the past, the Lamar wolves had used that trail, but I did not see them during the times I was on it. I sometimes searched for wolves just north of the park entrance at Gardiner, Montana, and twice saw the Eight Mile pack there.

Some days I drove east into Wyoming to look for the Hoodoo wolves. Those trips gave me a chance to see what their territory looked like, but each time I was over there I had to be careful, for wolf-hunting season had started on September 15. The Wyoming zone just east of the park had a quota of four at that time. It had been eight when 754 and 06 were shot the previous fall. I never did spot the Hoodoos. If I had spotted any, I would have left right away to avoid the risk of tipping off people to their location.

On one of those trips, a friend who knew the region showed me where 06 had been shot the previous December, less than a mile from the main highway. 754 had been killed the month before a few miles from there. The closeness of those two sites was another indication that 06 had likely gone back there to look for 754.

The shutdown ended late on October 15 and the park reopened the following morning. That day I saw seven Canyon pack wolves in Hayden Valley, including the alpha female, the wolf described earlier with the striking white coat.

Scott Frazier and John Potter returned to the park for their annual ceremony to bless the wolves. During the ceremony they prayed that the wolves would stay in the park rather than stray across the border and into the wolf-hunting zone.

Things were getting back to normal from the park closure and visitors were coming back to Yellowstone. One day several of us were helping new people see the Junction wolves through our scopes, and one woman said, "Everyone is so friendly here!"

I HAD NOT seen 755 since September 1. He had been alone that day. On October 21, we saw him up high above Lamar Valley on Specimen Ridge. He was with a black female from the Junction Butte pack. She had been collared in March 2013 and given the number 889. She had been born a year before the Junction Butte pack formed in 2012, which meant she was originally from the Mollie's pack. 755 had now been with two Mollie's females since the death of 06: 759 and 889.

I thought about how 755 would be seen by female wolves. He was a mature male with a lot of experience being an alpha and father. Compared with unproven younger males,

755 was a good choice for females looking to find a reliable mate. The *Bachelor* reality-TV show would have signed him up immediately.

AT THE END of the month, I checked my records and saw that Middle Gray had last been seen on October 10. When the shutdown ended, we all expected to see her, along with the other Lamar wolves, but she was not spotted through the end of the month. With her gone, the Lamars were down to just the two alphas, 925 and 926.

I saw 925 and 926 for the first time after the shutdown on November 19. They had a fresh bull elk kill east of the den forest, but the carcass was close to the road. Traffic and people there were keeping the wolves from feeding. I called the law enforcement rangers and we dragged the bull off to the south so the wolves would feel comfortable going to it.

By early December, we had twelve inches of snow on the ground in Silver Gate and it was getting down to minus 29 Fahrenheit (–34 Celsius) out in Lamar Valley. This was my fifteenth winter in Yellowstone, but I still had not adjusted to the cold weather. All the winters I had worked in desert parks in my earlier career, such as Death Valley, Joshua Tree, and Big Bend, had taken away most of the resistance to cold I had developed while growing up in New England.

WE FOUND 755 and his new mate 889 near Tower Junction one morning. 889 was leading and she had her nose to the ground as she followed a scent trail. There were hundreds of elk nearby, but 889 ignored them and concentrated on following one particular scent. That trail took the wolf through thousands of other elk tracks, but she stayed with the ones made by the elk she was interested in. 755 followed 889, and I sensed he understood she was on the trail of a likely target.

889 suddenly stopped, looked forward, then lowered her head into a stalking position. 755 joined her and stared in the same direction, toward a small group of conifers. The wolves must have been thinking the elk they were tracking was in those trees.

The female ran forward and 755 raced along with her. I lost them for a few moments in that stand of trees, then saw a cow elk running out of the forest with both wolves right behind her. The elk was running slowly and both wolves were rapidly catching up to her. As the wolves reached the cow, 755 was out in front. He grabbed a hind leg. She kicked back at him, hitting him several times, but he maintained his grip despite the blows. Then 889 jumped up and bit into the cow's throat.

The elk just stood there, no longer fighting back or even struggling to break free. She must have been in very poor health and had no strength left. The cow was so tall that 889 had to stand on her hind legs to keep holding on to the throat. Then the elk collapsed. She managed to get up but fell once more, and that was it. 889 held on and within a few minutes the cow was dead.

As I thought about that sequence, I concluded that 889 had found one set of elk tracks among thousands that had a scent she associated with sickness or poor health. She focused on that scent trail, ignoring all other scent trails, and it led directly to that cow. That was impressive. 889 seemed to be an especially intelligent wolf as well as a good hunter.

This gave me hope that she and 755 would succeed in finding a territory and starting a family. It was looking like 755 had finally found the right female to start over with.

Soon after that sighting, on December 14, I heard that a wolf had been shot and wounded just north of the park. I worried that 755's string of bad luck was continuing, because 755 and 889 were in that area at the time. On Christmas Day, I found 755 with 889 on Junction Butte, just east of Tower. The female held her front right leg off the ground when she walked. She bedded by 755 and licked that leg and her right shoulder. It turned out that she was the wolf shot on the fourteenth.

Doug McLaughlin stayed with the pair when I left to check on other wolves. He later told me 755 and 889 got up and affectionately licked each other's muzzles. The pair was near a new elk carcass the next morning. 755 must have killed it by himself, since the female could not have run fast enough to catch up with the elk.

We saw 755 by himself on the north side of Lamar Valley on the twenty-ninth. Later we spotted 889 a few miles behind him. She followed a set of wolf tracks east that must have been his. 889 did not put weight on her injured leg as she traveled. We could not see any sign of a break, so the shot may have gone through her flesh without hitting a bone. We lost her heading toward 755. Signal checks indicated the two were together up in that section of the valley for the rest of the month. I knew how resilient wolves can be and hoped that 889 would recover from her gunshot wound. 755 had been through enough tragedies in his life, and I wanted to see him succeed in finding a mate and once again father and raise pups.

On January 1, 2014, I saw 755 with 889 in Lamar Valley. As she followed him, she dragged her injured leg while wading through deep snow. Later she used that leg when she had to bound through snow that was even deeper. I guessed that must have been very painful for her. A few days later, a class at the Yellowstone Institute saw 889 chasing a group of elk, using three of her legs. 755 pursued a different herd. He caught up with a cow elk, grabbed her by the throat, and wrestled her to the ground. 889 ran over and helped him finish her off. That was impressive for a wolf who had recently been shot.

The Comeback of Wolf 870

Serious injuries often slow female wolves like 889 down, but I am continually amazed at how they refuse to give up and keep on doing what they need to do to survive. In early 2013, Doug Smith darted the gray Junction alpha female and collared her. She would now be known as wolf 870. At that time, the pack had seven adults and two surviving pups, a black and a gray. An adult gray male soon joined the group and we assumed he had come from the Blacktail pack, for he was accepted by alpha male Puff, who also was a Blacktail. The Junction pack's territory, at Slough Creek, made them neighbors to the smaller Lamar pack.

Two days later, Puff and 870 got in a mating tie. We later determined that during the breeding, the female injured her neck. Laurie Lyman was with me when we saw the tie and she recalled that a pup had jumped up on the back of 870. That and the weight of Puff caused her to collapse. Perhaps her neck got twisted around as she fell and that caused the injury.

I watched her soon after that incident. 870 got up from a bedded position and tried to follow Puff when he walked off, but she collapsed after a few steps. Puff came back and stood by her, an action I took to mean he was concerned about her. She soon got up and followed him but was unsteady on her feet.

As I studied her through my spotting scope, I saw that her head hung down. It seemed painful for her to hold it up. Expert dog behaviorist Kirsty Peake studied 870 and concluded she had a pinched nerve in her spine that caused pain in her neck and likely down her front legs. At times she limped on her front left leg, and I saw her licking her legs like they were hurting her.

Those injuries soon caused 870 to lose her alpha status. Another female known as Ragged Tail took over that position. They were likely sisters from the Mollie's pack. I saw confirmation of that realignment when 870 approached the other adults while wagging her tail. Ragged Tail went to her, jumped on her back in a dominant posture, and pinned her. 870 did not fight back. I thought of the concern Puff had shown 870 and how that contrasted with Ragged Tail taking advantage of her sister's disability.

870 seemed to accept the new hierarchy and her lower status well. I noticed that the other adult

females in the Junction pack treated her like she was out of competition for the alpha position and rarely dominated her. Even Ragged Tail eased off on her. 870 took on the role of watching over the pack's two pups. She frequently stayed with them when the rest of the adults took off on a hunt. That helped 870, for it gave her more time to recuperate from her injury.

When the Junctions had a group howl, I saw that 870 would lift her head and join in, despite the pain she likely endured by putting her neck in that position. After the howl, her head would sag down once more.

When 870 did travel with the pack, she was usually last in line. She frequently had to stop to bed down and rest, then would get up, walk as far as she could, and rest again. Eventually she would fall way behind the others, but she would keep her pattern of resting and pushing on and eventually catch up with them. I guessed that those long marches were hard on her neck and she had to stop when the pain and exhaustion got to be too much for her. When she walked, her head was usually hanging low. I became tremendously impressed with how 870 was coping with adversity. There was no quit in her.

As 870 was slowly recovering from her neck injury, I saw that she and alpha male Puff continued to have a special attachment to each other, despite 870's lower status in the female hierarchy. That did not seem to matter to him. They played together and bedded down side by side. I did not see Puff play with the new alpha female, Ragged Tail. He frequently ignored her and sometimes lunged and snapped at her when she tried to interact with him, things he never did to 870. This seemed to be a case where a female retained the affection of the alpha male even after she lost her alpha status.

In the spring of 2013, Ragged Tail denned west of Slough Creek, up high on a slope. The other Junction adults, including 870, worked hard to support the new alpha female and her litter of four gray pups. I remember one long hunting trip that took those adults far from the pups. When they were on the return route, the healthy adults easily trotted up a high ridge and rejoined the pups. 870 struggled up that slope, far behind the others, but kept going forward, one slow uphill step after another, with her head hanging low, likely in terrible pain.

I finally lost sight of her going to where the pups would be. She was not their mother, just a lowly pack member, but 870 seemed determined to help care for them. I was especially impressed that 870 was working so hard, despite her injury and apparent pain, to help the female who had supplanted her as the alpha female. In human terms, she was a team player. 870's dedication to her pack was repaid in an especially touching manner. One of those gray pups was seen giving her a piece of meat.

By late April, 870 had recovered to the point where she was chasing grizzlies from carcasses and teasing the adult male wolves in her family. She did double scent marking with Puff, mainly when Ragged Tail was not around.

We did not often see the Junction pack in the summer months, but in August I saw Ragged Tail do a scent mark at a spot. Puff was nearby but he did a scent mark at a different place. 870 was with them and she marked over his site. That day, Puff and 870 played together once more. This was more proof of their bond.

In September I saw that at times 870 was out in front of the pack in lead position. She also pinned one of the adult females. By the end of that month, she was pinning two different females.

On October 27, I found the Junction wolves and saw that 870 and Ragged Tail were both in the group. 870 seemed to be fully recovered from her injury. The following day, Ragged Tail was gone, as was a Junction adult male. I could see that 870 was once again alpha female. We never found out what had happened during the previous night. I thought about it a lot and eventually guessed that Ragged Tail saw that 870 was back to full strength. If that was the case, perhaps she chose to leave the family with the male and set up her own pack rather than risk getting into a losing confrontation with 870.

In mid-November, I was watching the Junction wolves in the Hellroaring Creek area when several vehicles pulled in the lot. It was a crew from MTV who were shooting a reality show about the singer Kesha. They asked if they could film me talking to her about wolves. She came over and I told her how 870 had been injured the previous winter and lost her alpha female status but had later recovered and regained her position. Then I suggested she write a song about 870. I saw one of the crew members a few days later and he told me that Kesha did write a song partly based on 870.

Whenever I think of 870, I picture her trudging up that steep slope toward the family's den, way behind all the other adults, moving forward, never giving up. And she did that to help out the female who had overthrown her.

PART V

2014

18

Quest for
an Alpha Female

O N FEBRUARY 6, we had the coldest day I had ever
experienced in Yellowstone or anywhere else. Several
of us had stopped to watch the Lamar wolves, 925
and 926, up at the den forest. One man had a thermometer
that registered a temperature of minus 55 Fahrenheit (–48
Celsius). The wolves were comfortably resting and seemed
unaffected by the extreme cold. I was not as tough as them
and had to go back into my car to warm up.

The motto of the Wolf Project in the early years was "Get
the data regardless of the cost." That extreme cold and the
need to get up at 3:15 a.m. on the longer summer days were
part of the cost of studying wolves. In later years, a new government regulation prohibited us from working outside in
extremely cold weather.

Soon after that day, I saw 926 leaving Soda Butte Creek
with a big beaver in her mouth. Its tail was missing, and I

took that to mean the wolf had already eaten it. The tail is mostly stored fat, so it would be a delicacy. I lost her carrying the beaver uphill to the den forest. I went back to the creek and found a patch of snow with a lot of blood on it. Wolf tracks were there, along with a lot of sticks partially eaten by a beaver. 926 must have surprised the beaver and grabbed it before it could slip back into the water. An adult beaver can weigh thirty to sixty pounds, so it would provide several meals for 926 and her mate.

As I watched the Lamar pair in late February, I saw how playful they were. Male 925 went to 926 and cavorted with her. He did play bows and romped around her. Both wolves frolicked off together, then she turned and jumped around him. After more playing, he chased her. It looked like they had a strong pair bond, and that signaled he likely would be a good provider for her. I was impressed by 926. She had taken over the alpha female position, had a strong bond with her mate, and was proving herself to be a good hunter.

I was now frequently getting signals from the Lamar alpha pair toward the den forest. 926 had been born there in 2011, and it looked like she was going to carry on the tradition of having her pups in the same den forest where 06 and Middle Gray had denned. That was also where her grandmother, wolf 472, had been born in 2000 to 926's great-grandmother, wolf 40. Earlier Druid litters had been born there going back to 1997.

IN MARCH, I saw that 926 was pregnant. There were times when her signals indicated she had gone underground to prepare for giving birth. She was about to be three years old and would be a first-time mother, but 926 had two years of experience helping her mother and older sister raise their pups, so I felt she was well trained.

Over the next few weeks, 926 continued to stay close to the den forest. By early April, her pregnancy was in its late stages and her belly looked huge. I saw her run at four big bull elk, but she soon slowed down to a walk and just sniffed at their scent trail. It must have been too uncomfortable for her to run any significant distance.

On April 8, I saw 926 get three regurgitations from 925. The following day, she bedded down and just watched as the male tried to get a bull elk by himself. He failed. It was not like 926 to pass up a hunt, but her advanced pregnancy prevented her from helping her mate.

926 seemed to be in her den on April 27, since I could not get her signal. I thought it was likely she had had her pups that day. Because of the thick trees surrounding her den, I figured I would not see her for a while.

THE LAMAR ALPHAS were not the only pair I was monitoring during those months. I also wanted to see how 755 and his new mate 889 were doing. In late February, I saw 755 on Specimen Ridge. 889 and a Junction black female were following him. They were with him the next day, and 889 did a lot of flirting with him. She playfully jumped around him, pawed at his side, then pressed herself against his chest. During a later sighting, I once again saw 889 wagging her tail and jumping around 755. In response he did a play bow and romped around her. The wound on 889 caused by the gunshot seemed mostly healed now, but she still limped at times. The black female was gone, and I wondered if 889 had chased her away.

Later in March, I saw 755 with 889 in Lamar Valley a mile or so from the den forest. When 889 got up, she looked pregnant. I was happy to see her bulging belly. It seemed 755 had finally succeeded in starting over and would soon be raising a new generation of pups. We monitored 889's travels to try to figure out where she would den. 889's signals soon indicated she was denning in a thick forest a few miles west of Tower Junction.

But soon after her likely due date, we started to see 889 traveling with 755 well away from her den site. I reluctantly concluded her pups had not survived. Maybe her gunshot wound and the stress associated with it were too much for her body and caused her pregnancy to fail in the late stages. That must have been a bitter thing for 755 to deal with. He had lost his brother and his original mate, 06, to human hunters. Then he had to leave his family to find a new female. He got 759 pregnant, but she died before having her pups. After that, at least six females showed interest in him. He ended up with 889, paired off with her, and got her pregnant. She seemed to be denning, but no pups survived. That meant that for two years in a row 755 had lost his opportunity to raise pups. He was so set upon by tragedy that 755 seemed to be the wolf version of Job in the Old Testament. I wondered how he could keep going after all those terrible setbacks.

I soon noticed that 755 and 889 were mostly apart from each other. 889 was frequently traveling on the road and getting near cars and people. I felt 755 was too wary of humans to be comfortable with that, a normal trait for a wild wolf. It looked like their relationship was over. He was going to have to continue his quest to find the right mate and start a new family. It had taken 06 several years to find 755. The life stories of wolves like 06 and 755 show that finding the right mate can be a long and difficult task with many failures and setbacks, just like it is for us.

IN CONTRAST, THE Lamar alphas 925 and 926 seemed to be doing well. The male repeatedly left the den on hunts, then came back to feed the mother wolf. I drove over to the Yellowstone Institute, where I knew that a grammar school from Thermopolis, Wyoming, was taking the Park Service's Expedition Yellowstone class. After bringing the class, teachers, and parents out to see 925, I overheard one of the mothers in the group say, "It almost makes me cry, that's how happy I am." Thinking about what she said gave me more appreciation for the impact on people when they see a wild wolf in the park.

It got warm that day. 925 found a dead bison calf and fed on it. He still had most of his thick winter coat, so he later walked off to a nearby snow patch and bedded down on it to cool off. Later, Laurie Lyman saw 925 carry half of the calf toward 926's den, about four miles to the east. That much meat would keep her full for a couple of days.

On May 7, ten days after 926 had her pups, I saw her and 925 chasing an elk downhill from the den forest. That elk escaped, but the next morning, they had a new carcass south of the road. 926 came down and fed on it. I had earlier been concerned that a pack with only two young adults might have trouble feeding and caring for a litter of pups, but so far 925 and 926 were doing a masterful job. 926 was young, but she had been trained by one of the best mother wolves we ever had. In turn, 06 had been trained by her mother, Agate alpha female 472, and she had been mentored by Druid alpha female 42. That is a superior line of females.

From the first day I had seen wolves in Yellowstone back in early May of 1995, I would go home at the end of the day and in the evening write up detailed field notes on what had happened. On May 10, 2014, nineteen years later, I wrote my ten thousandth page of wolf observations.

I GOT A good look at 926 in early June when she was feeding on a bull elk carcass. I could see that her milk glands were very full. 925 was bedded uphill from the carcass and letting her have the choicest parts of the elk. That was a good indication that 926 had chosen the right mate.

Every summer back then, I met up with a group from the Texas Children's Hospital. The adults in the group were a mixture of parents and former patients of the hospital who had fully recovered from cancer. The children all had cancer but were doing well enough to make the trip. I would show them wolves and then give them a talk. Like most kids I spoke with, they loved stories about our famous mother wolves, such as 42, 06, and now 926.

July 10 was a big day. That evening, I was watching the den forest from the Footbridge parking lot and saw seven Lamar pups in a meadow. Six were black and one was gray. I lost them going into some thick trees, but other people later saw the pups following 926 around. The sighting was an emotional event for all of us. After all the challenging times the Lamar Canyon pack had gone through in the last eighteen months, including the deaths of 754 and 06 and the departure of 755, those pups meant that 926, with the help of 925, was successfully carrying on what her mother had started.

Seeing 926 with her pups caused me to think about her father. Back in mid-May, I had heard that 755 was being seen in Hayden Valley, about twenty-eight miles south of Tower Junction. That was the territory of the Canyon pack, the new group I wrote about earlier. Their alpha male was from Mollie's and their alpha female was the daughter of the Hayden alpha pair who had been killed by the Mollie's wolves. Her coat was now mostly white, like her mother's had been. She had an adult daughter with a light gray coat. I immediately thought this daughter could be a good match for 755. I was rooting for him to find the right female and get to raise another litter of pups. After all the difficult times he had been through, he deserved some luck.

On July 21, we had a major sighting. I drove down to Hayden Valley and saw 755 with that young, light-gray Canyon female. They traveled north, with her in the lead, and I lost them going into a forest. The following day, I saw the pair flirting with each other. She went to him with a wagging tail and jumped on his rear end. Later she licked his face as he wagged his tail. 755 had spent the year and a half since the death of 06 looking for a suitable partner. I had a feeling that this female was the one he was destined to be with. He did pair off with her, but since it was long past the mating season, they would have to wait until next year to have pups.

IN MID-AUGUST, LONGTIME wolf watcher Kathie Lynch saw 926 trying to move her pups across the road to a new rendezvous site a mile southeast of the den forest. I joined Kathie and saw 926 holding a bone in her mouth. The pups chased her and tried to grab the bone from their mother. She occasionally put it down, then grabbed it and rushed off as the pups approached. This showed that 926 was using the bone as a lure to get the litter to follow her. Years earlier, I had seen Druid alpha female 42 use the same trick with a stick while trying to move her pups to the Chalcedony Creek rendezvous site.

When 926 got the pups where she wanted them, she bedded down, and they played among themselves. Soon 925 came in from the west. The pups ran over, and he gave them two regurgitations. Later 926 played with a black pup. In the coming days, I would continue to see her doing a lot of playing with her pups, as well as licking them. We consistently had a count of six pups rather than seven and concluded that one of the blacks had been lost.

The family stayed at that rendezvous site during the following weeks and I saw the pups playing together a lot. They looked healthy and seemed to be developing well. 926 continued to regularly join in on the games. She was a good mother to her first set of pups, just as her mother, 06, had taken good care of her first litter back in 2010. 926 and her family left that rendezvous site in late August and went up the Lamar River. By early September, they were at the Chalcedony Creek rendezvous site, the centrally located meadow their Druid ancestors began using in 1997.

The Lamar wolves were mostly out of sight during October. Late in the month, the pack reappeared in the valley. They chased a bull elk and 925 got a holding bite on his rear end. 926 got in front of the bull and he charged at her with a lowered head. The big bull might have gored her to death, but 925's grip held him back and he missed 926. Then the big bull bucked up and down and kicked back at 925. He hit the wolf twice but 925 held on despite the blows.

926 joined her mate and bit into the elk's hindquarters. The six pups were excitedly surrounding the bull but seemed too afraid to make contact. Then the alphas pulled him down in a patch of thick sage. The pups ran in and bit at the struggling bull. He soon was dead. That hunt would have been a prime learning experience for the pups. They witnessed how effective their father was in holding on to the bull and preventing him from escaping, then saw their mother join 925 and help take the elk down.

When I worked in Denali National Park, I often helped take care of the park's sled dogs. Sandy Kogl, the longtime kennel manager, told me that the younger dogs learned best by watching the older dogs perform their duties. A new dog would be put in a sled harness next to a veteran dog and quickly figure out how to properly act on the team. It works the same way with wolf pups watching older wolves.

We continued to see all six pups with the Lamar alphas throughout November. One day a grizzly came near the pack. A black pup went toward it and all the other wolves followed. They charged forward and surrounded the bear. 925 nipped it on the rear end, then bit it two more times. The wolves harassed the bear for a while longer, then lost interest and walked off.

A few days later, what seemed to be the same grizzly casually walked toward the pack. When the bear reached the wolves, some of the pups wagged their tails, like they were seeing an old friend. It walked through the wolf family and continued on. The wolves traveled with the grizzly, and it was often positioned in the middle of the pack. It did not bother the wolves and none of them harassed the bear. It looked like they had worked out a nonaggression treaty.

On the eighth, I reached a goal I had set for myself years earlier. I went out in the early morning for the 5,264th consecutive day, close to fourteen and a half years. That doubled Cal Ripken Jr.'s record of playing in 2,632 consecutive Major League Baseball games. If I could keep it up through next June, I would make it to the fifteen-year mark of going out every day. I had been going out so many days in a row that it was hard to keep track of what day of the week it was. I often had to ask people such questions as: Was it Thursday or Friday?

TWELVE MOLLIE'S WOLVES were on the north side of Lamar Valley on December 4, including six pups and their new uncollared gray alpha male. He had joined the pack in late 2012, about seven months after the Mollie's tried to raid 06's den. The Mollie's howled and we heard the Junction wolves howling back from south of the road. In response the Mollie's moved off to the north, indicating the new male and the alpha female had chosen to avoid a confrontation with the Junctions. It was a different pack now that violent 686 was no longer the leader.

The next morning, the Lamar wolves howled. The Mollie's and some of the Junctions howled back. A Junction black pup got separated from its pack and seemed to be confused by all the howling. It went toward the Lamars, probably thinking they were its family. I saw 926 staring at the pup. After more howling, the Lamar wolves charged at the pup with 926 leading. The pup now knew that this was not its family and ran away.

Since 926 had heard the Mollie's howling, she probably thought this pup was one of theirs. She had a legitimate grudge against the rival pack, as they had tried to attack her family at their den a few years earlier. 926 reached the pup and pulled it down. The pup broke free but she caught and pinned it. 926 nipped at the young wolf but the bites did not seem serious. The pup squirmed away and ran off. She chased it and could have easily tackled the pup again and killed it, but let it go.

The Mollie's wolves soon left the valley without causing any problems for 926's family. Those events seemed to confirm that the multigenerational feud between the Mollie's and Druid genetic lines had ended.

Four other wolves were in Lamar Valley that day. They were part of the Prospect Peak pack, which had been denning to the west, at Blacktail Plateau. In a few months, the Prospect wolves would have a fateful encounter with 926's family.

THAT SUMMER, I received regular reports that 755 and the Canyon female were still together in Hayden Valley. In late August, I saw them trying to get an elk calf. It ran into the Yellowstone River. The female waded out and tried to grab it but failed. The calf came out of the water and ran off. 755 chased after it at full speed. After a lengthy pursuit, the calf turned around and raced back to the river. 755 had already run a long distance but was still going all out and gaining on it. The calf ran into a shallow section of the river and the wolf grabbed a hind leg. As the calf waded into deeper water, 755 lost his grip and waded back to shore.

I studied him through my high-magnification scope and saw that he was not breathing hard. 755 was over six years old now. The average life span of Yellowstone wolves is four to five, so he was doing well for his age. The female came over and affectionately bumped into his side. Then she did a scent mark. He came right over and marked the site. Later he did a scent mark and she marked over the spot. After that, she playfully jumped up on his back and he did the same with her.

In September, I saw that the female's coat was now nearly white, the same tone as her mother and grandmother. She saw 755 coming toward her and ran to him wagging her tail enthusiastically. They greeted and he wagged his tail at a fast rate as well. The female bumped against him and he bumped his body against hers. They looked happy, like two playful dogs who were best friends. All of us who knew 755's story and all the tragedies he had gone through were so happy for him.

THE SHOOTING OF 754 and 06, along with the disruption their deaths caused in the Lamar Canyon pack, had a big impact on the visibility of wolves in Lamar Valley to park visitors. I went over my records and found that for the three-month summer tourist season (June, July, and August) during 2011 and 2012, the last two years of 06's life, we saw wolves on 80 percent of those days in Lamar. The next two summers, we saw wolves in Lamar on only 60 percent of the days. We also saw far more wolves in the earlier period compared with the later one.

Tourism is a major part of the economy of Montana and Wyoming, and Yellowstone is the main attraction in the region. When a 2005 survey was done by University of Montana professor John Duffield on why people came to the park, the top two reasons listed were wanting to see grizzlies and wolves. The number of people coming to Yellowstone to see wolves was estimated in 2005 dollars to bring in $35.5 million annually to the local economy. With inflation and increased visitation (Yellowstone now gets over four million visitors a year), I calculated that dollar figure would now be about $70 million. Most of that money goes to local businesses such as motels, restaurants, gas stations, and wildlife tour companies. All those companies employ local residents. A reduction in wolf sightings as a result of park wolves being shot outside Yellowstone impacted the owners and employees of those companies in a negative way.

The Matriarch

Wolves lead lives that are full of challenges. I have always admired how they continue on despite all the obstacles in their way. Of all the female wolves I have known in Alaska and Yellowstone, the one who had to endure the most adversity was wolf 42. She was captured as a pup with her mother and two sisters in southern British Columbia in early 1996, brought down to Yellowstone, and released into Lamar Valley that spring, when she was about a year old. Her mother was designated wolf 39, a gray sister wolf 40, and a black sister wolf 41. There was no adult male in the family, so a big lone gray, wolf 38, was added to the group prior to release.

40 turned out to have a violent personality. When she was a young adult, she drove off her mother and later did the same to her sister 41. Then she directed her aggression to 42 and constantly beat her up. When alpha male 38 died in the fall of 1997, wolf 21 joined the pack as their new alpha male. His personality was similar to 42's in that they both treated other pack members well and used cooperation rather than violence to accomplish things. From my observations, I felt that 21 was far more compatible with 42 than he was with 40.

In the spring of 1998 and again in 1999, 42 denned but none of her pups survived. Evidence strongly suggested that 40 killed both of her sister's litters. I often thought of the bitter anguish that 42 must have gone through after losing all of her pups in those two years, something made worse by the fact that it was her sister who had killed them.

42 continued to live her life as best she could with the continued aggression from 40. I could see that she regularly helped out the younger adult females in the pack, and this gave her some allies. In the spring of 2000, 42 denned five miles away from 40's den site

and two of the young females in the pack attended her there. Soon after 42 had her pups, I saw 40 approaching her den, which was located in a forest. I could not see what happened when 40 arrived but suspected that her intent was once again to kill her sister's pups.

The next morning, I found 40 by the side of the road. Her gray coat was drenched in blood and her body had so many bite marks on it that several wolves must have attacked her. She soon died of her wounds.

As I thought about 40, I felt she had tried to kill 42's pups the previous evening, but for the first time 42 had stood up to her sister and tried to stop her. 42 did not have the aggressive streak that characterized her sister. She had likely been losing the fight and in danger of being killed herself. At that point, I think her female allies jumped into the fray, on the side of the wolf who had treated them so well over the years. With it now being three against one, 40 would not have stood much of a chance.

Since I found 40 alive, I figured that when the other females defeated 40, they must have stepped back and allowed her to escape. But she did not survive her injuries and blood loss. The ultimate test of 42's character took place soon after that. Druid alpha male 21 traveled from 40's den to the den where 42 had her young pups and brought her back to 40's litter. He was desperate for help with the now-motherless pups. Since that den was in a thick forest, we could not see what 42 was doing with the pups. Was she caring for them, or was she killing them as 40 had killed her pups the previous two years?

42 went back to her den and one by one carried her pups to 40's den. Then she got two other Druid mother wolves to bring their pups to that central location. When we got a total count of twenty-one pups later that spring, we realized that 42 and the other mothers had cared for 40's litter as well as their own. Twenty of the Druid pups survived to the end of the year, a much higher percentage than the previous year when 40 was the alpha female. The following summer, under 42's benevolent leadership, the Druids numbered thirty-eight, the largest wolf pack ever recorded anywhere in the world. She converted what had been a dysfunctional pack to a superbly well-run, cooperative enterprise. One of her many gifts was organization. If she had been born a woman, she could have been the supreme allied commander in World War II.

It was during those years of watching 42 that it became clear to me that the alpha female is the true leader of a wolf pack, not the alpha male. I saw that 42 and other alpha females have an agenda and are forward-looking. They make the major decisions throughout the year, such as where to den and later which rendezvous sites to move the pups to. The survival of the pups and the long-term viability of the family depend on the wisdom of those decisions.

Male wolves like 21 are very respectful of females and seem to have no problem letting them lead the pack. I think that starts when males are young pups and learn that the mother wolf is the boss of all things. I often thought that adult males don't fully understand where pups come from and perhaps regard a female's power to create them as something like our concept of magic.

My favorite memory of 42 was from an incident that took place in 2002 when the Druids got into a fight with the Geode Creek pack. It was six of their adults against three Druids: 21, 42, and a pup. Because of 21's fighting prowess, his side was winning the battle. The conflict seemed to be over and 42 moved off, but then the Geodes came back and attacked the Druid pup. It ran to 21 and the rival wolves chased it. A big gray male was out in front. He bypassed the pup and attacked 21.

As 21 and the big Geode male fought, the pup got away. 21 broke off the fight and ran after the pup to protect it. All six of the Geode wolves pursued 21 and were just about to pounce on him as they all went out of sight into some trees. I pictured the Geode pack ganging up on 21 as he fought to protect that pup. Several minutes went by and I got increasingly worried as I imagined that they were killing 21. Then all six of the Geode wolves ran out of those trees and fled back uphill. They were running in a desperate attempt to stay ahead of a large group of Druids who were pursuing them. Soon, apparently satisfied that the Geodes were leaving the area, the Druids stopped and came together for a rally that I took to be a victory celebration. The wolves excitedly jumped up on each other as they wagged their tails. I counted thirteen wolves. That number included 42, but not 21.

Fearful that he had been killed, I desperately scanned the general area for 21 and still could not spot him. I later found 42 bedded down next to a big black wolf. It took me a few moments to realize it was 21. He was standing protectively beside his longtime mate. His fur was bloody and disheveled, but he had survived the gang attack. 21 had risked his life many times to protect 42 and the rest of his family, and this incident showed that she was just as willing to risk her life to save him. I think the other Druid wolves had learned to trust her judgment, and because of that they followed her as she charged to the site where the Geodes were attacking 21. That rescue mission was her finest moment.

As 42 and 21 grew old together, they seemed to become even more attached to each other, like a human couple who are totally devoted to each other. After 42 died at nearly nine years of age, a long life for a wild wolf, 21 was never the same. A few months later, he seemed to be losing his will to keep on living. 21 walked away from his family and died in a meadow that he and 42 had often visited together.

42 experienced a lot of hardship and tragedy during her life, but she had her partnership with 21 and the support of the other females in the family. With their help she triumphed over adversity. The era when 42 led the Druid pack now seems to have been a Camelot-like golden age for Yellowstone wolves.

PART VI

2015

19

Hard Times for 926

IN EARLY JANUARY, I watched Lamar alpha female 926 do a lot of playing with her six pups. She chased and wrestled with them, then the pups paired off and played with each other. 926 never knew her great-grandfather, Druid alpha male 21, but she took after him in the way she spent so much time playing with her pups.

Later a black pup dug a deep hole in the snow and came out with a big bison leg bone in its mouth. The hole in the snow looked to be about eighteen inches deep. The pup must have gotten the scent of the bone through that much snow.

On January 12, we had the twentieth anniversary of the arrival of the three wolf packs from Alberta to start the Yellowstone Wolf Reintroduction Project. By that time, the reintroduction was universally regarded as the most successful wildlife restoration project ever undertaken.

I GOT WORD that 755 and the light gray female had been seen together in Hayden Valley in late January. They had been a pair for over six months. The February mating season was about to start, and we hoped that in a few months we would get to see 755 raising a new set of pups. In April, I got a report that 755 and his female had been seen feeding on a bison carcass in Hayden Valley. After that, the pair was spotted hanging out at a den site where there was fresh digging. I knew that den site well, for this female had been born there. The Wolf Project decided to call the new group the Wapiti Lake pack. *Wapiti* is a Native American word for "elk." I drove down to Hayden Valley in early July and saw the Wapiti mother wolf bedded in the pack's rendezvous site. 755 walked over to her and two little black pups followed him. I went back a few days later and saw that he had fathered four pups.

At the end of my book on wolf 21, I asked the question "Can a wolf feel joy and happiness?" Watching 755 with those four pups and thinking about all the setbacks he had endured, I would definitely say yes. Whatever difficulties and challenges might later come his way, 755 was doing what he was born to do: be a father raising newborn sons and daughters with a female partner at his side.

FOR 926, 2015 started out as a normal year. She and 925 first mated on February 21. Two days later, the pack was feeding on a bull elk carcass. 925 tried to breed with 926 again, but she was more interested in playing with the pups. Hours later he finally succeeded in getting into a tie with her. After a few minutes, she seemed restless and moved toward the carcass, dragging him along. 926 reached the bull elk and began feeding. The male was positioned facing the opposite direction, unable to feed, so he stoically maintained the tie for a few minutes longer. Then the pair broke apart.

I like to play the straight man while doing talks. When I would tell that story and get to the part where 926 was eating while mating, I would pause, act forgetful, then ask the people gathered around me what the word was that described doing two things at the same time. It would always be a woman who yelled out, "She was multitasking!"

At the end of February, the Lamar wolves traveled way out of their territory to the west, about twenty-two miles from their den forest. That was almost as far as 21 had gone in days when the Druids controlled a vast superterritory. 926's family was much smaller and had only two adults in it, along with the pups from her first litter. They were taking a big chance in traveling that far from home. On March 2, they were safely back at the den.

Three days later, the Lamars were at Slough Creek, scavenging at an old bison carcass. The six pups were not quite yearlings but old enough to help their parents hunt. The pack went farther west with 926 leading. After traveling six miles, the wolves chased some elk. The herd split up and 926 and one of the pups went after a subgroup of three cows. Those elk ran to the top of a basalt cliff. They stood their ground there, facing away from the cliff. 926 stayed there and bedded down close to them while the pup left. The rest of the pack were near another elk herd.

When I was watching the main group of wolves, Laurie Lyman saw 926 approach a cow who stood on the very edge of the drop-off, grab one of her hind legs, and pull her down, despite being only a quarter of the size of her opponent. The elk got up but was unsteady on her wounded leg. As 926 approached, the cow stepped back a few inches too far and fell off the cliff. Laurie then saw 926 running down the side of the cliff toward where the cow had landed. Based on that report, I scanned the base of the cliff and saw 926 was feeding on the carcass.

She soon went back uphill and reunited with the rest of her family. I saw the pups greeting her. One of them must have gotten the scent of the fresh carcass on her, for it followed her scent trail downhill and found the cow elk. I was impressed with how intelligently that pup had acted. Soon after that, the rest of the pack, including 926, joined the pup and all of them fed on the elk.

THE PROSPECT PEAK wolves, the pack that had been denning at Blacktail Plateau, now numbered fourteen. They had become an aggressive pack and killed two wolves in January. 763 was the alpha male. Another large male looked like his identical brother, so we called him Twin. 965, a big collared gray male, was also in the pack.

On March 6, I saw twelve members of the Prospect Peak pack on a bison carcass at Slough Creek. I drove west and spotted the eight Lamar wolves heading that way from their elk carcass. They traveled through a pass on the ridge just west of Slough Creek and when they came out the other side, they picked up the scent of the Prospect wolves. They ran along the scent trail with alpha male 925 in the lead. 926 stayed at the rear, behind the six pups. Suddenly she turned around and fled back toward the pass. She must have seen the large pack down at the creek. Her pups were well trained, for they immediately ran off, following their mother.

But the alpha male did not move. 925 stood there calmly, in front of the pass, as twelve Prospect wolves charged uphill directly at him. The other members of his family were now out of sight, so the rival wolves could see only him. Just as they were about to reach him, 925 ran off, but not toward the pass. He ran the other way, toward the creek. All the other wolves chased him. I realized that 925 was leading them away from his mate and pups.

925 ran as fast as he could but the Prospects caught up, attacked, and pulled him down. I lost sight of him as the wolves clustered around him. The attack seemed to go on forever, but it probably lasted about four or five minutes. Several of the Lamar pups appeared on the ridge to the west and howled. All the Prospect wolves left 925 and ran toward the sound of the howls. When they failed to find any Lamar pups, they raced back to where they had attacked 925.

I swung my scope back to that site, expecting to see the lifeless body of 925. I could not spot him but figured the thick sagebrush blocked my view. The Prospects ran in and sniffed around. From their frenzied movements I could tell they could not find him. That meant that he had miraculously survived the brutal attack and escaped.

A black Prospect pup found 925's scent trail and ran with his nose to the ground, like a bloodhound on the trail of an escaped convict. It reached the creek and ran south, still following the scent trail. I scanned ahead and saw 925 running slowly. He was hunched up and obviously severely injured. His left hind leg seemed to be especially damaged.

The Prospect pup caught up with 925 and bit into that leg. Instantly the Lamar male spun around and lunged at it. Now terrified, the pup let go and ran off. 925 stood there briefly, then continued moving down along the creek. I soon lost him behind a hill. In the next few minutes, we spotted all six Lamar pups scattered throughout the area. Now we needed to find 926 and see if she was all right. The Lamar pups came together and went back toward Lamar Valley.

I did a signal check and picked up 926's signal to the west. I hiked back to the car and drove in that direction. A mile down the road, the signals from 925 and 926 were both to the north and seemed to be coming from the same place. That was a region of rolling hills covered with thick sagebrush. I had last seen 925 moving that way. It was now getting dark, so I headed home. On my last check, 925's signal was in normal mode, meaning he was likely alive.

On the way, I thought about what had happened. Being pregnant, 926 did exactly the right thing in running off to the west when she realized the bigger rival pack was about to attack. Her six pups also did the proper thing by following her. The mother and some of the pups would have been caught and killed if 925 had not stood his ground, in the direct path of the twelve oncoming Prospect wolves. When he ran off to the east, the other wolves could not resist chasing him. As they ran east, 926 and the pups were getting farther away to the west.

I WENT OUT early the next morning and got signals from 926 and collared pup 967 toward the den forest. Doug Smith was doing a flight and he called down to say 925's signal was coming from where I had picked up the wolf's signal the previous evening. It was a normal signal rather than the mortality one. The sagebrush there prevented Doug from spotting the wolf. I then heard the Lamar wolves howling from the den forest, likely trying to contact 925. Late that day, Wolf Project volunteer Emil McCain checked on 925 and got a mortality signal.

The next morning, 926 and all six pups were back at the den. I drove west and saw ravens and magpies flying into the sage patch where 925 had died. I then thought about how late on the day of the attack I had gotten signals from both 925 and 926 from that site.

I pictured him lying there alone, in great pain, likely knowing he was dying, then looking up and seeing 926 coming to him. It must have been an intensely emotional moment for him, for he now would have known that his sacrifice had worked. He had saved his mate's life by drawing the attacking party of wolves to him. She would have greeted him, then probably licked his face and his many wounds. After that, 926 would have had to leave so she could find her six pups and take them home. I guessed that she now understood she was going to have to look after her family without him.

I felt 06 had picked the right two males to help her form a pack, males she could rely on in a dangerous crisis such as the den raid. Now events proved that her daughter had also picked the right male, for he died heroically protecting 926 and her pups.

20

926's Triumph

I SAW 926 AND the six pups on March 9 in Lamar Valley. They traveled west and I later saw them at Slough Creek. 926 led the family to where the Prospect wolves had been feeding at the bison carcass, then took the pups to where 925 had been attacked three days earlier. The seven wolves sniffed around for a long time. All of them seemed subdued, and I would like to think that the pups understood what had happened there.

Seeing the family at that site caused me to have more empathy for 926. She had endured so many tragedies in her life. Her mother and uncle were both shot and killed within a few weeks of each other, then her father left the pack to seek out a new mate. She paired off with a male who was devoted to her, but he was killed by rival wolves. 926 was a single mother desperately trying to keep her six young sons and daughters safe in a very dangerous world.

The Lamar wolves were back at their den on March 12. Five days later, the family scavenged on two old carcasses.

One was a bison who had died the previous fall. I saw a pup chewing on one of the horns. Another pup gnawed on the skull. That indicated the pack had not had a fresh kill recently. I studied 926 and saw that she was visibly pregnant. If she did not eat better, her pregnancy would be in danger.

The following day, I saw 926 feeding on a mule deer she must have just killed. When her belly was full, she buried several portions of it in the snow. The pups also fed on the deer. Her mate had died just a few days earlier but 926 was carrying on, doing what she needed to do to support her family and get ready for the new litter of pups.

On March 21, we had the twentieth anniversary of the release of the Crystal Creek wolves from their pen. I had kept track of how many days I had been out in the park to look for wolves since that release date, and it came to 88 percent of the days in those twenty years.

The following morning, I got signals from Prospect gray male 965 in Lamar Valley. The signals were coming from just west of the den forest. We saw 926 and her pups up on the slope high above that forest. Later we spotted three of the big Prospect males on that slope: 965, Twin, and an uncollared black we called Mottled Black. Another black male known as Dark Black was bedded south of the road. He howled and the other Prospect wolves howled back from the north.

926 and the six pups were still up above the den forest. Later, at the end of the day, three Lamar pups were seen east of the den forest. They were going east, and the three Prospect males were following them. 926 and the other pups had stayed back near the den forest. That meant that some of the wolves who had killed 925 were now between 926 and three

of her pups. This was an extraordinarily dangerous situation for a single mother. Should she stay with the three pups still with her or leave them and attempt to rescue the other three?

I WENT BACK out early the next morning and saw a startling sight in Round Prairie, the meadow five miles east of the den forest. 926 was with the male known as Twin and they seemed comfortable with each other. During the night, she must have realized the arrival of the four Prospect males presented her with an opportunity. She desperately needed a new alpha male to help her with her current pups and with her new litter. I pictured her taking the risky decision to approach the biggest and most dominant male in the group in a friendly manner, wagging her tail. Her plan apparently worked, for they were now acting like friends.

To the north I saw the three other Prospect males. They were looking toward 926 and Twin. That pair howled and the three males howled back. The two wolves crossed the road to the north and soon linked up with the other males. 926 flirted with 965. Then she did a double scent mark with Twin. We had seen that male wolves found 06 irresistible, and now it seemed that her daughter had inherited the same type of allure. It had been only seventeen days since these males had killed her mate, and now 926 had converted them to her side.

The six Lamar pups were nearby and appeared to be scared of the new males, which was understandable after what had happened to their father. Twin led the group of adults toward the pups. Soon 926 took over the lead position. The four males seemed willing to be friendly to the pups.

The adults reached one of the female pups and the big males greeted her in an affable manner.

We saw 926 with the four males over the next three days but no pups were with them. For part of that time, signals from 926 and 965 came from the den forest, so they could have been with some of the pups there. I saw 926 flirt and interact with each of the males and that seemed to be working to bond them to her. We later had occasional sightings of pups with the adults, but they still were wary of them.

926 and the Prospect wolves went to an old carcass and the four males stood by as they let her feed first. After I left, Kirsty Peake saw the group kill a badger. 926 fed on it and would not let the males join her. When she later walked off, they went over and fed on the leftovers. The sightings at the two carcasses showed that 926 was now the dominant wolf in the group, despite being the smallest and most recent member. In just a few days, she had converted the four big males who had killed her mate from her enemies to her support staff. That was an astonishing accomplishment, something no other Yellowstone female had ever done.

The group was at Slough Creek soon after that and the five wolves worked together to kill a big bull elk. By that time, I had not seen the six Lamar pups for several days. On March 29, Dan Stahler did a flight and saw five of them at Cache Creek, to the south of the den forest. The pups were at a bull elk carcass.

In late March, 926 was up at the den forest, and I could tell from losing her signal that she had gone underground. The following morning, she was with three of the Prospect males south of the den forest. Eleven Mollie's wolves were a few miles northwest of them. They howled and Twin, who was now the Lamar alpha male, howled back. I saw that he and the other Lamar wolves were bedded down and seemed confident in their ability to deal with the larger pack. The Mollie's wolves had a group howl and the four to the south howled in response.

I stayed until it got too dark to see. The Mollie's were still north of the road when I left. It appeared that they did not want to confront the smaller group of Lamar wolves to the south. This incident proved that 926 had chosen her new packmates well. Despite being significantly outnumbered, those three males stood their ground by her, even when they knew that a much larger pack was nearby and might attack them.

One of the regular wolf watchers is a man named Steve Maras. He is a big man who was a star basketball player in college. Later he worked as a bodyguard for people like Mick Jagger. I began to think of the Prospect males as 926's personal bodyguards.

AS WE GOT into early April, all six Lamar pups were based well away from 926 and the new males. Probably they were still traumatized by the death of their father. The pups likely connected the scent of the new males with the scent of the pack that had killed 925. They later left Lamar Valley and went west. By that time, they were classified as yearlings.

I saw that 926 looked very pregnant in early April. Her new pups would have 925 as their sire but would be raised by the Prospect males. By that time, 926 was spending a lot of time up at the den forest. On April 27, I got her signal up there, then completely lost it. As before, I took that to mean she had gone into the den. I saw her later that day and noticed that she had distended nipples and missing fur under her belly, signs her pups were about to be born.

Two days later, Twin and Mottled Black were south of the den forest, close to a fresh cow elk kill. In the evening, 926 fed on the elk, then went right back uphill to her den. Twin followed her up there. The next morning, the pair went back to the elk carcass and 926 fed while Twin walked around and waited. He paused to look in various directions, like he was watching for any threats to the female. She grabbed a big piece of meat and carried it up to the den. Only then did Twin feed. Then he followed her route north, looking like he was guarding her.

IN EARLY MAY, all four of the Prospect males were regularly attending 926 at her den. The only one with a collar was gray male 965. He, along with Mottled Black and Dark Black, continued to be subordinate to alpha male Twin. By the fifth, 926 had recovered well enough from having her pups that she could travel several miles away from the den.

A few days later, I heard that the Lamar yearlings, including collared male 967, had been seen about ten miles north of the park. That was a dangerous place for young, inexperienced wolves because in the fall there would be a legal wolf-hunting season there. Soon after that, Doug Smith called to tell me the remains of a black male wolf had been found by a hiker. He was likely one of 926's sons, for the other five yearlings had been spotted nearby. There would be an investigation of the incident, and we waited for the results.

I continued to see how dominant 926 was to the males. The pack was at a bison carcass one day but only 926 fed. Three of the Prospect males were bedded down nearby, patiently waiting their turn. The moment she walked off, they rushed over and fed. I heard that at another carcass 926 and Twin got into a dispute, and he ended up on his back in a defeated position with the much smaller female standing over him. After putting the big male in his place, 926 fed on the carcass. That was more proof that an alpha female wolf is dominant to the pack's alpha male.

A lot of military veterans come to the park to see the wolves. I often talk to them about their experiences, then think about how what I have learned from them might apply to wolves. I gradually came to realize that in a wolf pack the alpha female is like a commanding officer, while the alpha male is her executive officer, meaning he carries out her agenda. That is especially true during the spring, when the female is tied up at the den caring for her new pups. The alpha male serves her by protecting the site from grizzly bears and rival wolf packs. He also organizes and leads hunting parties and brings back food to the mother and her litter.

I checked on 926's weight when she was collared and found that it was 82 pounds. When Twin was later collared, he was 110, and 965 was 119. That meant the two males outweighed her by 34 percent and 45 percent. 926 may have been a small female, but she had a big attitude.

That explained why the males were afraid to crowd in by her when she was feeding: she would beat them up. 926 had a lot of 06's defiant spirit in her. I saw more proof of that on a later day when 926 spotted a big mountain lion on a freshly killed elk carcass and charged at it. The lion ran off and climbed up a tree to escape from the wolf. None of the males helped her during the chase. They all went to the kill and stuffed themselves while 926 dealt with the cat. I later thought that if that lion had turned around and looked into 926's eyes, it would have seen the same fierce green fire that Aldo Leopold's mother wolf had shown a century earlier.

ONE MORNING I saw a black yearling going up into the den forest. Later I identified it as one of the two Lamar female yearlings. She was traveling with Twin, and both wolves seemed comfortable with each other. The other black female yearling returned to the pack right after that. I figured that, being females, they had no reason to be afraid of the new male pack members. 926's sons would likely be more concerned about the new males being aggressive to them.

With the arrival of the females, the Lamar pack had seven adults: 926, the two black females, Twin, Mottled Black, Dark Black, and 965. The pack was now a powerful force. I thought about all the wolves I had known over my time in Yellowstone and how certain ones had stood out. Rose Creek alpha female 9 was one of the fourteen wolves brought down for the 1995 reintroduction. She gave birth to wolf 21, the second generation. He sired 472, who later had the 06 Female. Years later, 06 had 926, the fifth generation of wolves I had known in the park. Her pups were the sixth generation. One of the two female yearlings who returned to the pack would eventually become the next Lamar Canyon alpha female. Her pups would be the seventh generation.

On June 12, I reached my goal of completing fifteen years of getting up early every day and going out into the park to study wolves. That day I saw all seven of the adult Lamar · wolves.

21

The Meaning of the Word Hope

IT TURNED OUT that the dead wolf found north of the park had been illegally shot. Since the Lamar yearlings had been seen in that area and one was now missing, that wolf must have been one of them. Whoever killed him left the body to rot.

The illegal shooting of the Lamar wolf was a hard thing for those of us who knew his family. The pack had endured so many tragedies, starting with the shootings of 754 and 06, and then the loss of 925. The killing of that yearling was one death too many for me.

I began to doubt if the thousands of wolf talks I did for park visitors were having much of an impact. I believed in the power of stories about the Yellowstone wolves, especially of 8, 21, 42, 06, 925, and 926. But were those stories making any real difference? Was everything I and so many other people had done to share their stories in vain?

Then there was a day that seemed to be the worst one of my career. I had been asked to give a talk to some young schoolchildren who lived in a town with many residents known to be very anti-wolf. I had never had a problem when speaking in such situations, but this time was different.

After the teacher introduced me, but before I could start to talk, a kindergarten boy, probably five years old, spoke up and told me, "I know the man that shot that famous wolf!" The killing of the 06 Female was still a big story in the region, and he had to be talking about her.

I could not think of a proper response. As far as we knew, her shooting was done legally, so being a National Park Service employee in uniform, I could not voice a political opinion about wolf-hunting regulations outside of the park. Having been a five-year-old boy myself, I could understand how this boy would have a heroic image of the man who killed the famous wolf.

I figured I would just start telling stories of the wolves I had known in Yellowstone and hope they might have some impact on these kids, especially that boy. But before I could begin, he spoke up again. He looked straight at me and said, "My dad just bought a license to kill a wolf!" Now I was really stuck. I could not think of anything to say that would not make the situation worse.

There was only one possible way out. I could say, "I need to start my talk, so no more questions or comments." But just as I was about to voice that line, the boy spoke for a third time. His first two words were "I hope... ," then there was a brief pause. I guessed what he was about to say: "I hope he shoots a big one." Or it might be even worse: "I hope I get to kill a wolf someday."

The boy finished his sentence with the words "I hope he doesn't!"

As far as I knew, that boy had grown up with frequent negative comments about wolves from the adults around him. But something had happened that caused him to reject that view, and I think it was the story of 06's life. Perhaps it was his teacher giving a lesson on wolves. Maybe it was Bob Landis's documentary on 06. The film was a big hit that played frequently on television. He had likely seen it, perhaps many times. I felt that 06's story had connected with the boy and changed him, and he wanted to tell me that he hoped his father would not kill a wolf.

That moment had a deep impact on me. It gave me renewed hope in the power of stories about wolves to reach people and change them. And it led me to finishing up with the Park Service so I could devote myself to writing books about the Yellowstone wolves. For years they allowed me to watch their lives, to see their pups grow up, find mates, and start families through many generations. I saw them in their good times and hard times. Most importantly, I learned their stories. And it was time to write those stories down so I could share them with everyone. No one ever truly leaves us as long as we keep on telling their story.

EPILOGUE

A T THE END of July, I watched the den forest in Lamar Valley and saw five pups running after one of the Lamar female yearlings. She gave the pups two regurgitations, then bedded and watched the pups play together. They chased each other and had a three-way tug-of-war over a stick. Later one ran away from the others, went behind a tree, and seemed to be playing a game of hide-and-seek. It looked out from one side of the tree at the other pups, ducked back behind the trunk, then looked from the other side. After that, 926 and the other six adults howled, and all five pups joined in.

Seeing 926 with her five pups illustrated the great resiliency that female wolves have. She went through times of great loss when 754 and 06 were shot. But 926 pushed on, moved forward, and did what she needed to do: find a mate and have pups. That mate was killed by other wolves, but she soon recruited four new adult males into her family and had another set of pups. I felt that she never gave up on hoping for a better future. 926 had a purpose in life, and that meant she continued on regardless of any losses or tragedies.

As I watched 926's litter, I thought about how many pups had been born to daughters of 06. In the three springs since her death, there were sixteen in the Hoodoo pack and another fourteen in the Lamar Canyon family, a total of thirty granddaughters and grandsons.

John Potter, in his wolf blessing ceremony, said the spirits of the wolves live on up in the mountains. The most promi- nent mountain in Lamar Valley is Druid Peak. It is positioned directly above the forest where 42, 06, and now 926 raised their pups.

I would like to think that the spirits of those two earlier symbols of female empowerment abide on Druid Peak, a site where they could look down and watch 926 and her new lit- ter of pups. Every one of those five pups had within them a little bit of those alpha females who had played such a criti- cal role in restoring a thriving wolf population to its rightful place in Yellowstone National Park.

AUTHOR'S NOTE

I DONATE PROCEEDS FROM my wolf books to Yellowstone National Park's Wolf Project through the nonprofit organization Yellowstone Forever (www.yellowstone.org /wolf-project) and to other organizations that benefit wildlife, children, and health care, such as the Make-A-Wish Foundation, Defenders of Wildlife, Earthjustice, Wolves of the Rockies, the International Wolf Center, Ecology Project International, and the American Red Cross.

ACKNOWLEDGMENTS

S O MANY PEOPLE have helped me over the years with my wolf research in Yellowstone and with my book projects. Here are the ones that I especially want to thank.

Yellowstone Wolf Project staff: Colby Anton, Cheyenne Burnett, Kira Cassidy, Lizzie Cato, Ky and Lisa Koitzsch, Hans Martin, Matt Metz, Pete Mumford, Rebecca Raymond, Doug Smith, Dan Stahler, Erin Stahler, and Jeremy Sunder-Raj. Thanks also to Yellowstone National Park bear biologists Travis Wyman and Kerry Gunther.

Researchers: John and Mary Theberge and Robert Wayne.

Photographers: Joe Allen, Kira Cassidy, Dan and Cindy Hartman, Jimmy Jones, Kim Kaiser, Ray Laible, Peter Murray, Eilish Palmer, Jeremy SunderRaj, Julie Tasch, and Bob Weselmann.

The Yellowstone wolf-watcher community: Jeff Adams, Stacey Allen, Rick Bancroft, Shauna Baron, Jim and Joellyn Barton, Ron Blanchard, Brad Bulin, Diane Busch, Becky Cox, Chloe Fessler, Marlene Foard, Jim Halfpenny, Larry and Linda Hamilton, Bill Hamlin, Calvin and Lynette Johnston, Sian Jones, Bob Landis, Laurie Lyman, Kathie Lynch, MacNeil Lyons, Steve and Robin Maras, Doug McLaughlin,

Pauline Meelis, Kara Menzel, Kirsty and Alan Peake, Mark and Carol Rickman, Carl Swoboda, Linda Thurston, Nathan Varley, and Story Warren.

Greystone Books staff: Publisher Rob Sanders, my editor Jane Billinghurst, marketing director Megan Jones, copy editor Brian Lynch, proofreader Meg Yamamoto, designer Fiona Siu, and rights director Andrea Damiani.

REFERENCES

Blakeslee, Nate. 2017. *American Wolf: A True Story of Survival and Obsession in the West*. New York: Crown.

Buschman, Heather. 2020. "How Old Is Your Dog in Human Years?" Press release, UC San Diego News Center, July 2, 2020.

Cross, P. C., E. S. Almberg, et al. 2016. "Energetic costs of mange in wolves estimated from infrared thermography." *Ecology* 97 (8): 1938–1948.

Duffield, John W., Chris J. Neher, and David A. Patterson. 2008. "Wolf recovery in Yellowstone: Park visitor attitudes, expenditures, and economic impacts." *The George Wright Forum* 25(1): 13–19.

French, Brett. 2021. "Late Hunt Finds More Infected Deer in SW Montana." *Billings Gazette*, February 23, 2021.

Geremia, Chris, et al. 2019. "Migrating bison engineer the green wave." *Proceedings of the National Academy of Sciences of the United States of America* 116(51): 25707–27713.

Halfpenny, James C., Leo Leckie, and Shauna Baron. 2020. *Charting Yellowstone Wolves: 25th Anniversary Edition*. Gardiner, MT: A Naturalist's World.

Hebblewhite, Mark, as quoted by Patrick Reilly. 2019. "Study: Yellowstone Bison Nature's Lawn Mowers." *Billings Gazette*, November 20, 2019.

Kraker, Dan. 2020. "New Research From Northern Minnesota Shows Wolves Feed Berries to Their Young." *MPR News*, February 12, 2020.

Landis, Bob. 2014. *She Wolf* (documentary by Bob Landis on the 06 Female). Trailwood Films and Media.

Leonard, Jennifer A., Carles Vilà, and Robert K. Wayne. 2005. "Legacy lost: Genetic variability and population size of extirpated US grey wolves (*Canis lupus*)." *Molecular Ecology* 14(1): 9–17.

Leopold, Aldo. 1949. *A Sand County Almanac*. Oxford: Oxford University Press.

McIntyre, Rick. 1986. *Denali National Park: An Island in Time*. Santa Barbara, CA: Sequoia Communications.

McIntyre, Rick. 1995. *War Against the Wolf: America's Campaign to Exterminate the Wolf*. Stillwater, MN: Voyageur Press.

Schweber, Nate. 2012. "'Famous' Wolf Is Killed Outside Yellowstone." *New York Times*, December 8, 2012.

Schweizer, Rena M., et al. 2016. "Genetic subdivision and candidate genes under selection in North American grey wolves." *Molecular Ecology* 25(1): 380–402.

Smith, Douglas W., Daniel R. Stahler, and Daniel R. MacNulty, eds. 2020. *Yellowstone Wolves: Science and Discovery in the World's First National Park*. Chicago: University of Chicago Press.

University of Wisconsin–Madison. 2007. "How Well Do Dogs See at Night?" *ScienceDaily*, November 9, 2007. www.science daily.com/releases/2007/11/071108140336.htm.

Wang, Tina, et al. 2020. "Quantitative translation of dog-to-human aging by conserved remodeling of the DNA methylome." *Cell Systems* 11(2): 176–185.

Wild, Margaret A., et al. 2011. "The role of predation in disease control: A comparison of selective and nonselective removal on prion disease dynamics in deer." *Journal of Wildlife Diseases* 47(1): 78–93.

Wilkinson, Todd. 2019. "Wildlife Diseases: A Global Expert Takes Stock of Greater Yellowstone." *Mountain Journal*, May 20, 2019.

INDEX

)